Modeling Cost and Performance for Military Enlistment

Report of a Workshop

Bert F. Green, Jr., and Anne S. Mavor, Editors

Committee on Military Enlistment Standards
Commission on Behavioral and Social Sciences and Education
National Research Council

NATIONAL ACADEMY PRESS
Washington, D.C.   1994

The work of the Committee on Military Enlistment Standards is sponsored by the Office of the Assistant Secretary of Defense (Personnel and Readiness) and funded under Defense Supply Services Contract MDA903-90-C-0028. The views, opinions, and findings contained in this report are those of the author(s) and should not be construed as an official Department of Defense position, policy, or decision, unless so designated by other official documentation.

Library of Congress Catalog Card No. 94-65606
International Standard Book Number 0-309-05041-3

Additional copies of this report are available from:
National Academy Press, 2101 Constitution Avenue N.W., Washington, D.C. 20418

B315
Printed in the United States of America

Preface

The Joint-Service Job Performance Measurement/Enlistment Standards (JPM) Project was initiated over a decade ago in response to congressional concerns regarding the quality of the All-Volunteer Force. In its initial stages, the project's efforts were devoted to demonstrating the feasibility of using hands-on tests to measure the job performance of enlisted personnel. The purpose was to provide a criterion measure based on job performance that could be used to validate the selection test used by the military to screen applicants for enlisted service. This stage of the project resulted in the construction of valid job performance tests and the collection of data for a cross-section of military jobs.

In the past three years, the focus of the project has been on incorporating job performance into the development of a cost/performance trade-off model to be used by military manpower planners in making decisions about recruit quality goals. This model links recruit quality to job performance on one hand and recruit quality to personnel costs on the other. Understanding these linkages provides a clear rationale, based on performance and cost differences, for choosing applicants for military service.

In June 1993 a workshop was held to present the cost/performance trade-off model to military manpower analysts as a tool for planning and justifying various mixes of recruit quality. At the workshop, presentations were given by individuals involved in developing the model, by Service representatives who had either used or carefully reviewed the model, and by members of the Committee on Military Enlistment Standards.

This report provides a set of papers on which workshop presentations were based. The context for the workshop and the need for the cost/performance trade-off model are introduced by the first paper, which examines trends in the quality of military personnel from the beginning of the All-Volunteer Force to the year 2000 and beyond. Other papers discuss technical issues associated with the development of the various components of both cost and performance linkages and present applications of the fully developed model. Following opening remarks by W.S. Sellman, Director of Accession Policy, Office of the Secretary of Defense (Personnel and Readiness), Part I provides an overview of the enlistment standards project. Part II includes two papers on job performance measurement issues. They represent two approaches to generalizing performance results to jobs for which no performance data are available and to interpreting job performance scale scores. Part III includes two papers describing the cost/performance trade-off model and its applications.

The committee is indebted to the workshop presenters and participants for their contributions to the success of the workshop. In addition to the authors included in this volume, the committee would like to extend its thanks to William Carr, Gerald Laabs, Captain Gary Macomber, and Major James Thomas for their thoughtful comments on the use of the cost/performance trade-off model. We also gratefully acknowledge the support and encouragement of Dr. W.S. Sellman, Director of Accession Policy, and his staff, Lt. Col. Thomas Ulrich and Jane Arabian. Finally we wish to thank Carolyn Sax for her efforts in planning the workshop and preparing the manuscript for production.

Bert F. Green, Jr., Chair
Anne S. Mavor, Study Director
Committee on Military Enlistment Standards

Contents

vii

Modeling Cost
and
Performance for
Military Enlistment

Opening Remarks:
The Nexus Between Science and Policy

W.S. Sellman

Frequently, when people ask what I do, I tell them I live in the shadows between science and policy. My office, the Directorate for Accession Policy, is involved directly with public policy. In fact, we may be one of the few public policy offices within the Department of Defense (DoD). This is because the people with whom we deal—young men and women—are civilians. They are considering military service, but, at the time they make this decision, they are still civilians.

All other offices within the DoD with military personnel management responsibilities deal with Service members—people who already have entered service and taken the oath of appointment or enlistment. The rules that pertain to the two populations—civilians and military personnel—are fundamentally different. Because we deal with public policy and civilian youth, we have perspectives different from military personnel managers. Since many of us in Accession Policy are psychologists and we are involved with activities such as personnel testing, selection, and classification, we are sensitive to and attempt to comply with the procedures and rules established by our professional organizations. My personal philosophy is that you cannot set good policy without analysis, without data. Consequently, we attempt to define the various situations and issues we encounter so we can collect data that inform policies—hence, life in the shadows between science and policy.

One of the criticisms frequently directed at social science is that we spend a lot of time and money on research for which everybody already knows the answers and nobody wants. I believe the Joint Service Job Performance Measurement/Enlistment Standards (JPM) Project is a good example of the linkage between science and policy. It also illustrates how research can be conducted and then implemented in a way that actually changes practices and procedures.

JOB PERFORMANCE MEASUREMENT AND ENLISTMENT STANDARDS

The JPM Project had its origins in the misnorming of the DoD enlistment test. The version of the Armed Services Vocational Aptitude Battery (ASVAB) in use from 1976 to 1980 was miscalibrated. As a result, the Services enlisted hundreds of thousands of people who would not have met the enlistment standards, had the test been measuring ability accurately. When the department informed Congress about the misnorming, Congress was very concerned; however, there were two positive outcomes to this particular announcement.

The first was that Congress encouraged the department to conduct the Profile of American Youth study, in which we administered the ASVAB to a nationally representative sampling of young people to develop contemporary norms. Until that time, the norms in use tracked back to the 1944 mobilization population. Today, after 14 years, the 1980 Profile norms are starting to get old. We now are involved in planning new norming studies, because never again can unqualified personnel enter service because of a flaw in our testing system.

The second positive outcome of the miscalibration was creation of the JPM Project. Congress was surprised to learn that we validated the ASVAB and enlistment standards against performance in training. Consequently, Congress told us to link enlistment standards to job performance.

With this mandate, we initiated the JPM Project in summer 1980. There were some early studies that considered the cost trade-offs between aptitude and performance on job knowledge tests. As we gained momentum, we established a Joint Service working group and asked the Service personnel research laboratories to undertake projects to measure hands-on job performance. Ultimately, we sponsored the Committee on the Performance of Military Personnel to provide state-of-the-art scientific oversight.

When we started the job performance measurement research, recruit quality was low, in the aftermath of the ASVAB misnorming. In 1980, more than 50 percent of all Army recruits were in AFQT Category IV (percentiles 10 to 30), and only about 55 percent of new Army recruits were high school graduates. The department responded well to this personnel crisis, and with the change in administration, recruiting and advertising resources increased and we began to improve recruit quality. By the mid-

1980s, we were recruiting 90 percent high school graduates with the percentage of new recruits scoring in Category IV falling to about 5 percent.

RECRUIT QUALITY: ISSUE OF SCIENCE OR POLICY

As the job performance measurement research evolved, we developed hands-on performance tests for about 30 occupations, which covered between 25 and 30 percent of all enlisted personnel. This was not done without enormous effort on the part of the Services and a lot of money. When we started the project, we had two basic objectives. The first was to learn: Can we actually measure job performance? For years, industrial psychologists had contended that performance was the ultimate criterion for validating selection tests. There always had been reasons why people could not measure performance, and basically it came down to cost. Measuring job performance is a very expensive proposition. With the support of Congress and the department's effort to recover from the embarrassing misnorming episode, the money for developing hands-on performance tests became available.

The second objective was: If we can measure performance, can we develop procedures for linking enlistment standards to that performance? Because the Office of the Secretary of Defense (OSD) was involved in this effort, there was skepticism about our motivation by the Services. I said then, as I say today, OSD does not intend to set Service enlistment standards. We will not define them, nor will we enforce them. Rather, through a cooperative effort, we will develop procedures that will provide the Services a more scientific basis for establishing their standards.

Throughout the 1980s, recruit quality continued to improve, and it became clear that the kind of personnel management problems that we experienced at the beginning of the decade were rapidly evolving into entirely new challenges. Because the department's recruiting budget had reached astronomical levels—about $2.2 billion—the question by 1985, became: How much quality is enough? As a result, there were many questions from members of Congress about why the Services needed this level of quality, and why we should spend so much money to attain it.

In 1985, the department submitted a report to the House and Senate committees on Armed Services establishing military recruit quality requirements for the rest of the 1980s, and we were virtually defenseless to justify those requirements and the associated budgets. We could say that smart people perform better than less-smart people. We also could say high school graduates have more perseverance than nonhigh school graduates and stay in service longer. But if the questions were: What is the difference between having 90 and 95 percent high school graduates or between 60 and 70 percent recruits scoring in AFQT Categories I-IIIA (percentiles 50-99)?—we could not answer. Thus, the JPM Project began to take a different shape and the policy question was no longer: How much quality is enough? Rather, the question became: How much quality can we afford?

RECRUIT QUALITY, PERFORMANCE, AND COST

By the early 1990s, we moved into a new phase of the project, which was to use job performance information in enlistment planning. We began developing models that allowed us to link recruiting resources, quality, and job performance. Once we had the models, if Congress said, what happens if we cut your budget by 10 percent, we could respond, quality would go down by x percent and performance would go down by y percent. We also were in a better position to defend our budget requests.

As an aside, you have read in the press about Secretary of Defense Aspin's recent initiatives on readiness. Within OSD, readiness has become an important issue. Somehow, we must organize ourselves so we can measure readiness, and Mr. Aspin can counter charges that, with the serious cuts both in the size of the force and the Defense budget, the military again is like the hollow forces of the late 1970s.

You also have seen in the press recent articles about declines in recruit quality, with the usual alarmist statements by senior officials, particularly within the Army, about the fragility of the All-Volunteer Force. With all due respect to our Army brethren, this is congressional testimony time, and it is useful to make statements that will defend the budget. The truth is that recruit quality is down somewhat. However, if you consider recruit quality levels for the 1980s, at which time the Services, including the Army, said the force was excellent, the quality that we have recruited thus far in fiscal 1993 is still considerably above the average that we experienced throughout that decade.

In looking at predicted performance levels using the job performance model, we conclude that a floor of 90 percent recruits who are high school graduates and 60 percent scoring in AFQT Categories I-IIIA are benchmarks to which we should aspire. The levels of recruit quality that we are experiencing today still exceed those levels.

What you are seeing in the press, both from senior civilian leadership and from the military, is the seasonal dance we do with Congress, but you should not be unduly troubled that recruit quality is really on a downward spiral. Our latest statistics show that recruit quality is rebounding, even for the Army. In May 1993, the Army recruited 99 percent high school diploma graduates and 69 percent scoring in AFQT Categories I-IIIA.

DENOUEMENT

Today, the Job Performance Measurement Project is almost complete. For years, I have said that the project was the holy grail of industrial psychology—the pursuit of the ultimate criterion of job performance. I also believe that this effort and the Army's Project A have made major contributions to industrial psychology and to other professions. I receive numerous telephone calls from people outside industrial psychology who have obtained copies of the book by the Committee on the Performance of Military Personnel on performance in the workplace; they are impressed with potential applications to their professions. I also should tell you that the National Academy of Sciences' new Board on Testing and Assessment will provide counsel to a number of government organizations facing difficult and complex measurement problems. Most of these problems, in the early going, will deal with educational measurement issues, i.e., accountability, measurement of proficiency within and across schools, minimum curriculum and performance standards. I believe the work done on the Job Performance Measurement Project will assist the new board in its definition and solution of such measurement problems.

It is gratifying to look at the JPM Project over the last 14 years and realize that we have been able to sustain one research project for that period of time. Research comes and goes, but this project is unique because it existed over five administrations, both Democratic and Republican. It cost a great deal of money—in the vicinity of $40 million. Despite the skepticism and, in some cases, objections by various people, the research has gone forward and now is close to fruition. In fact, I view this workshop as the beginning of its completion.

APPRECIATION

As I look around the room, I see several people who have been with this project from the beginning, and I feel compelled to recognize them. First are Bert Green and Sandy Wigdor, whose scientific and policy contributions have been invaluable; they always have provided sage advice. My admiration for them is certainly not a secret. Then there is Dave Armor, who did the very first job performance measurement study in 1980. Dave has been with the project as a scientist, as a policy official within the department, and now as a member of the Committee on Military Enlistment Standards. Dave has been an inspiration to me in all roles. I respect enormously his analyses as well as his policy guidance. Dave, I thank you for your help, your counsel, and your support.

There also are representatives from the Services who have been with the project almost all the way—Larry Hanser and Jane Arabian, in the beginning with the Army Research Institute. Larry now is with the Rand Corporation, and I am pleased that Jane is in my office. Bill Strickland, Armstrong Laboratory, has viewed this research from several perspectives—as manpower staff officer, recruiter, and now director of human resource research for the Air Force. Jerry Laabs, Navy Personnel Research and

Development Center, also designed and conducted much of the work in measuring job performance of Navy ratings.

In addition, there are the chairs of the Job Performance Measurement Working Group, two of whom are here today: Dickie Harris, Human Resources Research Organization (HumRRO), and Tom Ulrich, Defense Manpower Data Center. Chuck Curran, also with HumRRO, who was the first chair, had planned to attend, but was recently taken ill. I have spoken with Chuck several times over the last several weeks. He is feeling better, and he asked that I send his regards and tell you how much he wanted to be here to see how the project had evolved since the early days. Finally, there is my old and dear friend Dick Hoshaw—Navy manpower policy wonk and humble implementor of research. Dick has been with the project from its inception, and his vision, wise counsel, and perceptive advice have been worthy.

Cost/Performance Trade-off Model

Where do we plan to go with the job performance measurement model? The project's first objective was to set enlistment standards, and I still intend to do this. Tomorrow, Rod McCloy of HumRRO will tell you about efforts to use the model for setting standards as a function of quality requirements. As I indicated earlier, I do not intend to tell the Services what their standards should be. What I will do is provide a methodology and encourage the Services to use it in establishing their standards. I hope the Services will be sufficiently persuaded by the quality of the research and the validity of the model that the standards-setting process will be compelling.

We also will use the model for planning and budgeting activities. As we develop the budget, we will run the model to see if budgets are realistic. If a Service is underfunded, we will ask the Service comptroller to reprogram money. We recently did this with the Navy, because of underfunding of its advertising program. If the Services are overfunded, I doubt we will try to cut the money from the budget, given the fact that the Congress is inclined to do that anyway. If Congress does try to cut the budget, we will use the model to defend our submissions. We also will do our best to persuade congressional staffers that the budget levels are reasonable and appropriate, and to try to get them to look elsewhere for cuts.

I am pleased that the National Academy of Sciences is hosting this workshop. I thank Anne Mavor and Carolyn Sax for their hard work in its planning and conduct. I also thank each of you for coming and hope that you will find the workshop to be a rewarding experience.

Part I:
The Context of the Enlistment
Standards Project

The Joint-Service Job Performance Measurement/Enlistment Standards (JPM) Project is one of the largest coordinated studies of job performance on record. Initiated by the Department of Defense (DoD) in 1980 and scheduled for completion in 1994, the JPM Project represents an investment of many millions of dollars and involved the participation of thousands of people. Phase I of the project, which concentrated on developing a variety of measures of job performance so that enlistment standards could be related to something close to actual performance on the job, included measurement specialists who designed the performance tests, local base personnel who provided logistical support for the data collection, and the more than 15,000 troops who supplied the performance data. In Phase II, econometricians worked with measurement specialists to develop a cost/performance trade-off model that incorporated the relationship between job performance and recruit quality on one side and the relationship between recruit quality and the cost of recruiting, training, and attrition on the other. This model provides a useful aid to accession policy planners who are responsible for deciding such questions as: How much quality do we need? or How much quality can we afford?

For the past 10 years, two committees of the National Research Council-National Academy of Sciences have served in an advisory capacity to the Department of Defense on the JPM Project. The Committee on the Performance of Military Personnel was formed in 1983 to provide an independent technical review of the research and measurement issues involved in (1) the development of hands-on job performance tests for jobs of first-term enlisted personnel, (2) the collection and analysis of data from test administrations, and (3) the linking of resulting performance scores to military enlistment standards as defined by scores on the Armed Forces Qualification Test (AFQT). In 1989, the Committee on Military Enlistment Standards was established to oversee the technical issues concerning the development of a cost/performance trade-off model for use in setting enlistment standards.

The sheer size of the JPM effort over that past 13 years ensures that the project will provide a wealth of raw material and guidance for the next generation of researchers in the field of human resource management, quite aside from its more immediate goal of improving the selection and classification of military enlisted personnel. The project's many achievements add in important ways to the understanding of personnel selection systems. Even its shortcomings are informative, for they point up the need for additional methodology and highlight the dilemma resulting from conflicting purposes that are inevitable in a project of this magnitude.

ORIGINS OF THE JPM PROJECT

In 1973, Congress abolished military conscription, and the military establishment was faced with the prospect of maintaining a competent active-duty military force on the basis of voluntary enlistment. Intense public debate accompanied the move to the All-Volunteer Force. Many feared that able volunteers would not sign up in sufficient numbers. Opponents warned that the national security would be weakened. Others were concerned on social and philosophical grounds that the burden of national defense would fall largely to those who would have most difficulty finding work in the civilian economy—minorities, the poor, and the undereducated (Fullinwider, 1983). With the matter of exemptions from the draft made moot by the shift to a volunteer force, military manpower policy came to revolve around issues of recruit quality and the high cost of attracting qualified personnel in the marketplace (Bowman et al., 1986).

Concern about the quality of the All-Volunteer Force reached a climax in 1980, when DoD informed Congress of an error in scoring the Armed Services Vocational Aptitude Battery (ASVAB), the test used throughout the military since 1976 to determine eligibility for enlistment. A mistake had been made in the formula for scaling scores to established norms, with the result that applicants in the lower ranges of the ability distribution were given inflated ASVAB scores.

The ASVAB includes 10 paper-and-pencil ability tests covering factors of verbal ability, mathematical ability, clerical speed, and technical knowledge. For enlistment purposes, the general aptitude of service applicants is assessed by a composite of the four ASVAB tests that make up the AFQT. For policy purposes the AFQT score scale is divided into five categories (Category III is frequently divided into IIIA (50-64) and IIIB (31-49)):

CATEGORY	AFQT SCORE RANGE
I	93 - 99
II	65 - 92
III	31 - 64
IV	10 - 30
V	1 - 9

The current enlistment standards and quality goals imposed by Congress for the entire armed forces are as follows. The legislated minimum standard for high school graduates is 10; in other words, those with scores in Category V are not eligible for military service. Since some military occupational specialties are more difficult than others, and the more difficult jobs go beyond the capabilities of lower-scoring recruits, it is necessary to enlist people in the upper score ranges. Legislation requires than no more than 20 percent of the enlistees be drawn from Category IV (score range 10-30). The misnorming of the ASVAB in 1976 led to enlistment of approximately 250,000 applicants between 1976 and 1980 who would not normally have been accepted (Office of the Assistant Secretary of Defense—Manpower, Reserve Affairs, and Logistics, 1980a, 1980b).

In response to the misnorming and to allay its own broader concerns about building an effective enlisted force solely with volunteers, DoD launched two major research projects to investigate the overall question of recruit quality. The first project, conducted in cooperation with the U.S. Department of Labor, administered the ASVAB to a nationally representative sample of young people between the ages of 18 and 23. This Profile of American Youth (Office of Assistant Secretary of Defense—Manpower, Reserve Affairs, and Logistics, 1982) permits comparisons between the vocational aptitude scores of military recruits and the test performance of a representative sample of their peers in the general population as of 1980. No longer do the test scores of today's recruits have to be interpreted with test data from the World War II era.

The profile provided important evidence to quell the worst fears about the quality of the All-Volunteer Force. The scores of enlistees for fiscal 1981 on the four subtests of the ASVAB that make up the AFQT were higher than those of the 1980 sample of American youth. In particular, the proportion of enlistees in the average range was considerably larger, and the proportion of enlisted personnel in the below-average range smaller, than in the general population. Although the results were reassuring, the weakness in the evidence was that quality was defined in terms of the aptitudes of recruits, not realized job performance—that is, in terms of inputs, not outputs. The relation between test scores and performance on the job was not established empirically, and thus DoD still could not satisfactorily answer the more difficult questions about the quality of the voluntary military service: How much quality is enough to ensure a competent military force? Given the need to compete in the marketplace for able recruits—using the lures of enlistment bonuses, high entry-level pay scales, and educational benefits—how much quality can the country afford?

In 1980, the assistant secretary of defense in charge of manpower and personnel affairs called on the Services to investigate the feasibility of measuring on-the-job performance and, using the measures, to link military enlistment standards to job performance. With the endorsement of the House and the Senate Committee on Armed Services, the Joint-Service Job Performance Measurement/Enlistment Standards Project, DoD's second major research project, got under way. The progress of this massive research effort is charted in an ongoing series of annual reports to Congress from the Office of the Assistant Secretary of Defense and the two-volume report of the Committee on the Performance of Military Personnel published in 1991.

Now, after more than a decade of research, empirical evidence has replaced assumptions about the efficacy of the ASVAB. The JPM Project has successfully measured the job proficiency of incumbents in a sample of military entry-level jobs. In the process, it has compared several types of measures and different approaches to test development. The performance measures provide a credible criterion against which to validate the ASVAB, and the ASVAB has been demonstrated to be a reasonably valid predictor of performance in entry-level military jobs.

Generalizations from the JPM results will take their place in the literature and lore of industrial and organizational psychology. Because of the superior measures of performance, constructed with a care normally reserved for standardized tests used as predictors, these results provide a solid base for general conclusions formerly based on less satisfactory criteria.

PROVIDING COST/PERFORMANCE TRADE-OFFS
IN LINKING ENLISTMENT STANDARDS
TO JOB PERFORMANCE

Phase I of the JPM Project demonstrated that reasonably high-quality measures of job performance can be developed, and that the relationships between these measures and the ASVAB are strong enough to justify its use in setting enlistment standards. But the human resource management problem is not solved by showing that recruits who score well on the ASVAB tend to score well on hands-on performance measures. High-quality personnel cost more to recruit, and the public purse is not bottomless. In order to make reasonable budgetary decisions, Congress needs to be able to balance performance gains attributable to selecting those with better-than-average scores on the ASVAB against the costs of recruiting, training, and retaining high-quality personnel. And to improve their control over performance in the enlisted ranks, DoD and the Services need to be able to make more empirically grounded projections of their personnel quality requirements.

The second phase of the JPM Project concentrated on the development of a cost/performance trade-off model to illuminate for policy makers the effects of alternative enlistment standards on performance and costs. The development of this model for setting enlistment standards has great potential relevance for accession policy. Until now, the standards-setting process has been largely based on an informal process of individual judgments and negotiations among the stakeholders. The manpower management models used by military planners for other purposes have simply assumed an appropriate enlistment standard or have used surrogates at quite some remove from job performance. With the JPM performance data incorporated into trade-off models, the models offer policy officials useful tools for estimating the probable effects on performance and/or costs of various scenarios— say a 10-percent reduction in recruiting budgets, a 20-percent reduction in force, or a downturn in the economy. The solutions provided by such models are not intended to and will not supplant the overarching judgment that policy officials must bring to bear, but they can challenge conventional assumptions and inject a solid core of empirical evidence into the decision process.

The full implications of the job performance measurement research for military policy makers—and for civilian sector employers—remain to be worked out in coming years. The JPM Project has produced a rich body of data and a wealth of methodological insights and advances. And, as important research efforts so frequently do, it has defined the challenges for the next generation of research on performance assessment.

WORKSHOP PRESENTATIONS

The workshop papers presented in this volume represent the culmination of the committee's activities and the JPM Project efforts. The purpose of the workshop and these papers was to provide military manpower planners and analysts with a description of the cost/performance trade-off model and concrete examples of its use for policy decision making. The workshop presentations and background papers address the following issues:

- Trends in military manpower quality: past present and future.

- Setting performance goals based on data from the JPM Project in comparison with alternative approaches to setting performance goals.
- Extending the performance equation from jobs for which hands-on performance data exist to jobs for which no performance data were collected.
- Understanding the underlying assumptions and variables comprising the cost/performance trade-off model.
- Showing how the cost/performance trade-off model can be used to examine the implications for recruit quality mix of budget reductions or changes in performance goals.

REFERENCES

Bowman, W., Little, R., and Sicilia, G., eds.
  1986   *The All-Volunteer Force After a Decade: Retrospect and Prospect.* Washington, D.C.: Pergamon-Brassey's.
Fullinwider, R., ed.
  1983   *Conscripts and Volunteers: Military Requirements, Social Justice, and the All-Volunteer Force.* Totowa, N.J.: Rowman and Allanheld.
Office of the Assistant Secretary of Defense (Manpower, Reserve Affairs, and Logistics)
  1980a   *Aptitude Testing of Recruits.* Report to the House Committee on Armed Services. Washington, D.C.: U.S. Department of Defense.
  1980b   *Implementation of New Armed Services Vocational Aptitude Battery and Actions to Improve the Enlistment Standards Process.* Report to the House and Senate Committee on Armed Services. Washington, D.C.: U.S. Department of Defense.
  1982   *Profile of American Youth: 1980 Nationwide Administration of the Armed Services Vocational Aptitude Battery.* Washington, D.C.: U.S. Department of Defense.

Military Manpower Quality: Past, Present, and Future

David J. Armor and Charles R. Roll, Jr.

U.S. military leaders have always believed that manpower quality—traditionally defined in terms of educational attainment and aptitude as measured by standardized tests—is just as important as quantity in determining force capability and readiness. The problem as we approached the 1980s was that the combination of a voluntary force, shrinking personnel budgets, and a youth population declining in size had raised serious doubts about whether sufficient quality could be maintained in the enlisted forces given the manpower requirements at that time.

Furthermore, the only way to maintain or raise enlisted recruit quality at that time was to increase personnel and recruiting budgets (i.e., salary, bonuses, education benefits, advertising), always a challenging proposition in Congress. The selling job was made all the more difficult because the military lacked credible models to show how much manpower quality was necessary and at what cost. Although each Service argued that quality requirements were based on empirical analysis, the fact is that there was no formal relationship between recruit quality and force capability or readiness, and no strong link between capability and costs. The unofficial justification for high-quality recruits, it seemed, was "the more quality the better."

The Joint-Service Job Performance Measurement/Enlistment Standards (JPM) Project aimed to remedy this problem by developing a formal manpower model that could help determine enlistment and quality standards by first linking quality to actual job performance and then linking performance

13

to manpower costs. By trading off performance and costs, the model would seek optimal quality standards given a variety of manpower constraints and conditions such as force size, compensation levels, and various external considerations such as unemployment rates.

By the time the JPM Project had produced an enlistment standards model, however, U.S. defense policies had changed drastically in response to world events. The end of the cold war reduced the national security threat, and the reduced threat led to reductions in force structure and manpower requirements, with corresponding decreases in numbers of recruits at all levels of quality. Moreover, the size of the youth population reached its nadir in 1993 and will begin rising modestly for at least the next 10 years.

With these changes in forces and demographics, the questions about quality in 1993 are vastly different from those in 1980. The problem today might be rephrased: Not how much quality is enough, but how much quality is too much? In 1980 quality was at all-time lows, while in 1993 quality is at all-time highs. Just as recruiting costs had to rise when the demand for high-quality recruits exceeded supply, the question today might be: Do recruiting costs fall when the supply of high quality exceeds demand? If the JPM enlistment standards model justifies higher recruiting costs during the 1980s when high quality was in shorter supply, will it show reduced recruiting costs during the 1990s if supply exceeds demand? What does the JPM model say, if anything, about future trends in quality, and in particular the relationship between quality requirements in the active and the reserve forces?

THE HISTORY OF THE QUALITY ISSUE

Manpower quality was not a major issue after World War II, when enlisted manpower requirements could be satisfied through a peacetime draft. The draft led to induction of a reasonable cross-section of young American men, although it did underrepresent certain categories of aptitude. Aptitude is defined here in terms of categories on the Armed Forces Qualification Test (AFQT), which is a subtest of the Armed Services Vocational Aptitude Battery (ASVAB). The highest levels of aptitude, AFQT Categories I and II (the 65th to the 100th percentiles), were often underrepresented because of college deferment policies in effect at various times during the draft era. The lowest level of aptitude, AFQT Category V (below the 10th percentile), was underrepresented because of longstanding policies that barred enlistment for this group.[1]

---

[1]In the Services, high quality is defined as recruits who are high school graduates and who achieve an AFQT score in the 50th percentile or above (Category I-IIIA).

Quality and the All-Volunteer Force

With one exception, quality did not become a major issue until the end of the draft and conversion to the All-Volunteer Force in 1973. The exception occurred in the mid-1960s, during the Johnson administration's war on poverty. At that time, more blacks volunteered for military service than were then enlisted—many of them with lower AFQT scores. The President ordered Project 100,000, the purpose of which was to improve job opportunities for youth with low socioeconomic status and minority youth by enlisting 100,000 Category IV applicants (10th to 30th percentile on the AFQT). As we shall see, this project played an important role in some of the early research on enlistment standards.

Manpower quantity and quality became an issue of great concern during the debate over the draft versus the All-Volunteer Force. Supporters of the draft argued that quality standards could not be maintained if enlisted recruiting was voluntary, because higher-quality youth would not volunteer in the face of more attractive college or job options, at least without making compensation so high as to make military budgets unaffordable. Supporters of the All-Volunteer Force argued that (1) adequate numbers of higher-quality personnel could be recruited with moderate increases in compensation and (2) no harm would come to the military even if the quality mix shifted downward somewhat, because many military jobs could be performed competently by Category IV enlistees. This latter view was adopted by the 1970 report of the influential Gates Commission, which had been appointed by the President to evaluate the viability of the All-Volunteer Force, and which recommended that the draft be replaced by the All-Volunteer Force.

During the early years of the All-Volunteer Force, the Gates Commission conclusions appeared to be vindicated. By all measures of quantity and quality, including AFQT scores, each of the Services was meeting staffing requirements and maintaining a quality mix comparable to that of the draft era. In fact, a comprehensive review by the Rand Corporation in 1977 concluded that the aptitude of recruits had actually increased, although there was some decrease in education levels (Cooper, 1977). The proportion of total Department of Defense (DoD) Category IV recruits fell from 19 percent in the latter draft years to about 6 percent in the early years of the All-Volunteer Force and from 24 to 11 percent in the Army. The percentage of non-high school graduates increased from 30 to 35 percent in DoD and from 33 to 44 percent in the Army.

This optimistic picture was shattered in 1980, when the assistant secretary of defense for manpower reported to Congress that the ASVAB in use since at least 1976 had been misnormed. The norming error caused AFQT percentile scores to be inflated, thereby causing recruit quality levels to be overstated. When the ASVAB norms were corrected and AFQT scores were recalculated, true quality levels changed dramatically. Instead of increasing quality, aptitude levels had deteriorated to their lowest levels since aptitude tests were adopted before World War II. As of 1980, when the norming errors were reported, Category IV personnel had risen to 35 percent for DoD as a whole, and to fully 50 percent for the Army.

Needless to say, there was a great deal of controversy and debate about this surprising turn of events, including calls for a return to the draft and claims that the All-Volunteer Force had failed. Those policy makers still committed to the concept of a volunteer force had a range of responses to the new quality crisis. At one extreme, the secretary of the Army attacked the validity of the ASVAB, arguing that quality standards were unnecessary in the first place and discriminated against minority citizens in the second. Others acknowledged the importance of quality standards and immediately called for increases in military pay and benefits in order to remedy the quality problem. It so happened that military pay had been falling behind civilian pay during the late 1970s, due to a combination of complacency and antimilitary sentiments left by the Vietnam War. The approach selected was to increase military pay and benefits.

The Search for Quality Standards

The response by Congress and both the Carter and Reagan administrations was to preserve the volunteer force but to embark on an extensive program to raise military pay and benefits sufficiently to remedy the quality shortfall. This outcome, while favored by most defense policy makers, was not without significant repercussions. Since defense spending is always problematic in Congress, the question now became: How much quality is needed and for what price? While almost everybody agreed that the quality levels in 1980 were too low, what constituted the "right" levels? Was the draft-era quality mix the right level? Could it be lower without sacrificing military capability? Should it be higher, given the increasing technological sophistication of weapons systems? Were draft-era quality levels affordable, or might we have to settle for lower manpower quality because of budget constraints?

As of 1980 there were no ready answers to these questions. There were serious data deficiencies and, more important, a lack of satisfactory tools for developing more rigorous validation methodologies. First, although each Service had enlistment quality standards at the time, those standards were based on the relationship between quality and several surrogate measures of military performance and capability, some of which had unestablished relationships with true military performance. Initially, most Services set enlistment quality standards by validating aptitude scores against training school outcomes, which at one time had relatively good relationships with

ASVAB scores. By 1980, the relationships between ASVAB scores and training results had weakened considerably, especially in the Army, where staffing requirements practically forced schools to graduate all trainees regardless of performance. Moreover, in the Army there was little relationship between ASVAB scores and such performance measures as time to promotion, supervisor ratings, and attrition rates and no data at all on the relationship between ASVAB and on-the-job performance.[2] In other words, in the Service with the most serious quality problem, there was little empirical basis to defend the argument that higher quality increased military capability by improving either training success or job performance.

Second, reliable data was just beginning to emerge concerning the cost of recruiting higher- (versus lower-) quality personnel, since the All-Volunteer Force—and free market competition between the military and civilian jobs—had been in existence for only six or seven years. Finally, no methodologies had been developed for validating enlistment standards by linking quality, job or training performance, and the cost of recruiting higher- quality personnel. Such validation methodologies were necessary for answering the critical question: How much quality is enough and for what cost?

The quality crisis led to several early experimental efforts to validate enlistment standards against job performance and costs. Most of this early work was carried out at the Rand Corporation, beginning with Rand's 1976 evaluation of the All-Volunteer Force. Analysis was carried out on data from an evaluation of Project 100,000, which had developed comprehensive and detailed hands-on performance tests for four Army jobs—armor crewman, vehicle repair, supply specialist, and cook (Vineberg and Taylor, 1972). These Project 100,000 job performance tests were unique in several respects, including: (1) use of job analysis to determine both critical and commonly performed tasks; (2) design of a performance test instrument for each task, in which each step of a task could be scored as performed correctly or incorrectly; and (3) a field testing situation in which job incumbents actually performed each task (with real equipment) and were scored by a former noncommissioned officer experienced in those jobs.

During 1980-1981 Rand carried out additional analyses on a second source of data, the Army Skill Qualification Test (SQT). The SQT, developed as an operational version of the job performance tests developed for Project 100,000, included three components: a hands-on performance test; a

---

[2]The Rand report by Cooper reported extensive analyses that showed virtually no relationship between ASVAB and attrition nor between ASVAB and specially designed supervisor rating measures (but education status did correlate significantly with both performance measures). This is not to say that there were no relationships between ASVAB and written performance tests, of which the Army Skill Qualification Test (SQT) was a good example (see Armor et al., 1982).

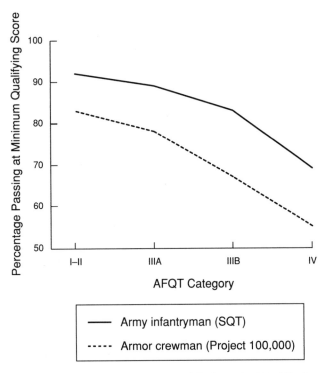

FIGURE 1 Job performance and AFQT (from the Rand Project 100,000 and SQT studies).

job knowledge test, and a certification component (e.g., firing range qualifications). Reasonably reliable SQT tests had been developed for several Army jobs, the first of which was infantryman. The Army began using the SQT as a basis for promotion to the higher grade levels, requiring a score of 60 percent (correct) to establish a minimum level of job proficiency.

The early Rand studies led to two preliminary conclusions. First, there was a substantial relationship between ASVAB scores and on-the-job performance tests for first-term enlistees, including the largest combat specialties. Figure 1 shows the relationship between AFQT and the percentage passing on-the-job performance tests for the two largest Army combat jobs. The on-the-job performance tests shown are the SQT for infantryman and the Project 100,000 test for armor crewman.[3] The relationship between

---

[3]An analysis of SQT scores showed that approximately 95 percent of noncommissioned officers were able to qualify at 60 percent for several jobs; using that criteria, a passing score on the Project 100,000 tests was set at 50 percent correct.

percentage passing and AFQT categories is quite strong. For armor crewmen, over 80 percent of Category I and II first-termers could pass the Project 100,000 performance test, compared with only about 55 percent of the Category IV personnel. Category IIIA also had high qualifying scores, but less than 70 percent of Category IIIB first-termers passed. For infantryman, only 70 percent of Category IV passed the SQT, compared with over 90 percent of Category I and II first-termers.[4]

The second conclusion was that optimal quality standards could be determined using a cost/performance trade-off model. Using a criterion of *qualified man-months*—the number of months of service by first-termers who remained in the military and who could pass the SQT—it was found that the cost per qualified man-month (including recruiting, training, and compensation costs) was much higher for the 1980 quality mix (which included 50 percent Category IV) than for alternative mixes with higher levels of quality. The model for infantryman showed an optimal cost-performance mix: about 18 percent Category IV, 30 percent nongraduates, and 35 percent Category I-IIIA high school graduates, the latter being considered by the Services to be high quality (Armor et al., 1982). An improved model was applied to four Army jobs in a later study (Fernandez and Garfinkle, 1985).

The quality crisis also generated a response by Congress. Although the analytic work was still under way, it was clear that the very low quality levels of the 1979 and 1980 recruits were going to be unacceptable by any standard or model. On the basis of discussions with the Services and with some input from those conducting formal analyses of quality requirements, in 1981 Congress passed legislation setting maximum limits of 20 percent Category IV and 35 percent nongraduates for new recruits. It also responded positively to calls for higher military pay, benefits, and recruiting budgets. Military pay was increased to be more competitive with civilian jobs, education benefits and enlistment bonuses were approved, and recruiting budgets expanded considerably to pay for more recruiters and national advertising campaigns.

Finally, the quality crisis inspired the JPM Project, which was initiated by the assistant secretary of defense for manpower in 1980 but formally mandated by the Defense Appropriations bill for fiscal 1983. The JPM Project was the first DoD effort to attempt to formally validate enlistment

---

[4]The SQT could not be administered until a person had been on the job for at least 3 months (and commanders could wait longer). The Project 100,000 tests were administered to enlistees with all levels of experience, including 1 to 3 months (and nearly half of armor crewman had less than 16 months of service). This probably accounts for the high passing scores on the SQT.

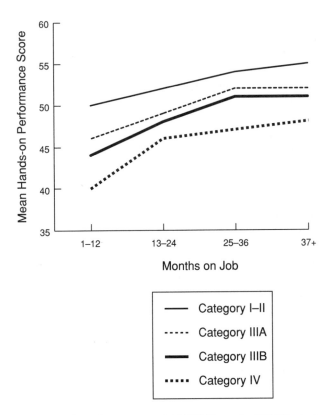

FIGURE 2 Job performance and AFQT (from the JPM Project, by AFQT and job experience).

quality standards against on-the-job performance criteria in all the Services. A Joint-Service working group was established, comprised of both policy and technical representatives from each of the Services. The working group established technical standards for hands-on performance tests and ASVAB validation requirements, and for the next 10 years the working group monitored the development and administration of hands-on job performance tests as the first step in the validation process.

Figure 2 presents a summary of the basic relationship between AFQT, job experience, and hands-on performance tests developed by the JPM Project for 25 jobs in all four Services (first-term enlistees only).[5] The relationship between AFQT and hands-on performance is not as strong as that found in

---

[5]This relationship was first reported to Congress in January 1989. See Office of the Assistant Secretary of Defense (FM&P), *Joint-Service Effects to Link Enlistment Standards to Job Performance: Recruit Quality and Military Readiness*, 1989.

the original Project 100,000 study, but it is consistent across all levels of experience for first-termers. The most important difference, that between Category I-II and Category IV personnel, amounts to about one-half of a standard deviation. Note that there is also a fairly small difference between Category IIIA and IIIB first-termers, who score about halfway between the highest and lowest categories. The experience relationship is also weaker than many analysts might have expected, but it is strongest for Category IV personnel. A Category IV first-termer with three years of service scores about the same as a Category IIIA personnel with one year of service. At the end of the first term, however, higher-aptitude personnel still score higher than persons with lower aptitudes, but the relationship is not as pronounced as at the beginning.

The fact that ASVAB scores are correlated with on-the-job performance measures provides considerable rationale for DoD policies that base eligibility for service on ASVAB test scores. Yet this relationship by itself does not provide a specific set of quality standards or a quality mix that answers the question: How much quality is enough, and for what cost? The answer to that question requires a cost/performance trade-off model, which was finally developed between 1990 and 1993 and is described in detail elsewhere in this volume.

QUALITY REQUIREMENTS AND TRENDS

Although there was no fully developed analytic methodology for validating enlistment standards against cost and performance criteria until the early 1990s, the Services nonetheless had to set and apply enlistment standards throughout the 1980s. The Services did have some empirical data to justify quality requirements, but for the most part they relied on the positive correlation between aptitudes and performance coupled with the traditional assumption that "more is better." Indeed, as will become clear from the quality trends described later, "more is better" can be inferred as the only consistent standard for enlisted quality between 1980 and 1993.

Quality Requirements, 1985-1989

In the 1985 Defense Authorization Act, the Services and DoD were required to review trends in enlisted quality and to establish quality requirements for 1985 to 1989. The study was motivated by congressional concerns that the high levels of quality recruited in 1984 might not be sustainable or affordable due to a number of factors, including economic recovery, falling unemployment, and the shrinking supply of youth.

Figure 3 presents the enlisted quality requirements established by each of the Services in this study. The figure shows three indicators that summa-

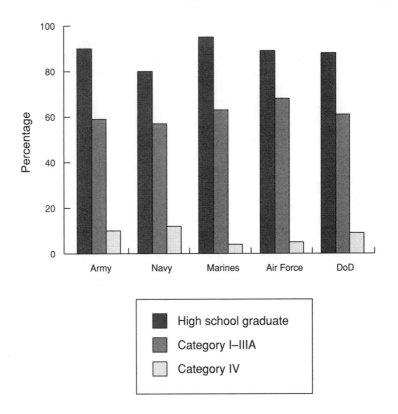

FIGURE 3   Quality requirements (for active enlisted accessions without prior service, fiscal 1985-1989). SOURCE: DoD report to Congress on defense manpower quality, May 1985.

rize enlisted quality standards for a Service as a whole: percentage high school diploma graduate; percentage AFQT Category I-IIIA (over the 50th percentile); and percentage AFQT Category IV (10th to 30th percentile).

The stated requirements for high school graduates range from 80 to 95 percent, with an average of 88 percent for DoD as a whole. The requirements for Category IV are very low by historical standards, ranging from 4 percent in the Marine Corps to 12 percent in the Navy and averaging 9 percent for DoD as a whole. The requirements for Category I-IIIA average 61 percent for DoD as a whole—very high by historical standards. This quality indicator runs about 5 points higher than the DoD definition for high quality, Category I-IIIA high school graduates, which was not defined in the DoD report. Interestingly and not surprisingly, these five-year requirements are very close to what each of the Services had recruited in 1984. In other words, the Services told Congress that the actual enlistment quality levels as of 1984 could suffice as requirements over the next five years.

As we shall see, actual enlistment quality did resemble these requirements between 1985 and 1989 but not after 1989. To our knowledge, neither the Services nor DoD conducted a formal quality requirements study or revised quality standards for later years, even though conditions changed dramatically with the force structure reductions beginning in 1990. Now that appropriate technical tools have been made available by the JPM Project, it would be quite easy to remedy this situation and establish new quality requirements for the new, smaller force structure of the 1990s.

Trends in Enlisted Accessions and Quality

One of the principal factors that determines the supply and the cost of higher-quality recruits is total staffing requirements as dictated by the force structure. Holding the quality mix and other external factors constant, a larger force means more high-quality enlistees, which in turn increases the marginal cost of recruiting additional high-quality people. Likewise, as the force shrinks, recruiting costs for a constant quality mix should also decline. This relationship is modified by the size of the youth population, with the cost of high quality increasing as the youth population declines.

Figure 4 plots the total number of active enlisted accessions without prior service between fiscal 1981 and 1992, with projections for fiscal 1993 to 2000 based on a 1.4 million active force as proposed by President Clinton. To indicate changes in the youth population, the figure also shows enlisted accessions as a percentage of men ages 18 to 21 in the general population, which accounts for about 70 percent of all new recruits.

The active force size of about 2.1 million during most of the 1980s led to a relatively constant number of about 300,000 nonprior-service enlisted accessions. Enlisted accessions declined to about 275,000 after Defense budget reductions in 1988 and 1989, and then to about 200,000 when President Bush's 25 percent force cut took effect in fiscal 1991. Assuming that the active force size falls to 1.4 million as proposed by President Clinton, the enlisted accession requirement will fall to about 180,000 starting in 1993.

The reduction in active force size has had a dramatic impact on the fraction of the youth population needed to meet the corresponding accession requirements. The greatest pressure on supply occurred during the mid-1980s, when accessions were stable and the youth population was declining, during which time the fraction of youth needed for accessions rose to about 4 percent. After 1989, however, the fraction of youth needed for requirements began falling sharply, and by the time the size of the youth population reached its lowest point in 1993, the drop in accession requirements

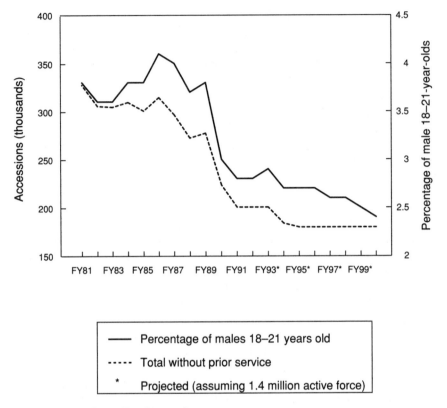

FIGURE 4   Active enlisted accessions.

lowered the fraction to less than 3 percent. Moreover, the youth population begins growing after 1993 (due to a mini baby boom during the late 1970s), which means the fraction of youth needed for recruiting by fiscal 2000 should reach a modern-day low of less than 2.5 percent.

Given the reduced demand for enlisted accessions, both supply and cost considerations become more favorable for higher-quality recruits. The Services could react to this more favorable recruiting climate in one of two ways: they could keep the quality mix constant and reduce recruiting costs, or they could keep recruiting costs relatively high and increase the number of higher-quality recruits. It appears that the second scenario is the one actually adopted, in spite of the stated quality requirements shown in Figure 3.

Actual trends in enlisted quality between 1980 and 1992 are shown in Figure 5. The combination of increased military pay, education benefits, enlistment bonuses, advertising, and other recruiting improvements led to

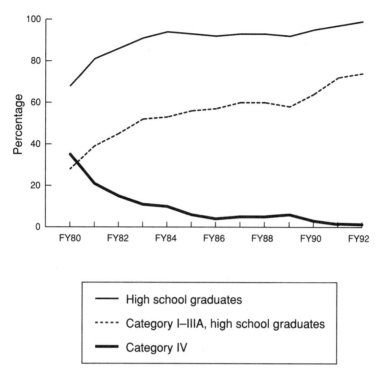

FIGURE 5 Trends in quality: education and aptitude indicators for active enlisted accessions.

dramatic increases in quality levels between 1980 (the low point in quality for the All-Volunteer Force) and 1985. For DoD as a whole, the percentage of high school graduates climbed from 65 to 90 percent, the percent Category IV declined from 35 to 5 percent, and the percentage of high-quality recruits—Category I-IIIA high school graduates—increased from 28 to 56 percent during this five-year period. These figures are unmatched not only in the years of the All-Volunteer Force, but also during the peacetime draft years after World War II.

The quality mix was relatively stable between 1985 and 1989, with few notable changes in either education or aptitude levels. In fact, the actual quality mix is quite close to the quality requirements shown in Figure 3, with the exception that the actual percentage of Category IV is about one-half of the stated requirements. Starting in 1990, however, and coinciding with the large force reductions after 1989, the quality levels again increased sharply. By fiscal 1992, remarkably, there were 1 percent non-high school graduates and only 1 percent Category IV personnel. Even more significant, the percentage of high-quality recruits grew to an unprecedented 75 percent.

There is very little basis in either military research or military policies since World War II to suggest that a capable military force requires three-fourths of its enlistees to be from the upper half of the national distribution of vocational and mental aptitudes, or that there should be virtually no one from the lowest one-third of the distribution. Although no official quality requirements were announced after 1985, the fact that the most recent quality mix is so much higher than the 1985-1989 official requirements—which were already at historic highs—suggests very strongly that the fundamental quality standard is "more is better" (if not "all is best")!

Quality Supply and Demand

The increasing quality levels after 1989 can be explained in several ways. Common to any explanation is the fact that total accession requirements have declined significantly. We would expect that the demand for military service would also decline, not only because of reduced recruiting emphasis but also because of the Persian Gulf War. One scenario that would explain increased quality levels is that the supply of military service has decreased more among lower-quality youth than among higher-quality youth, to the point that there are not enough lower-quality youth to meet the 1985 requirements. Another scenario is that there is excess supply at all levels, and the military services are simply choosing to maximize whatever quality is available.

We can evaluate the supply situation in part by examining trends in applicants for the active enlisted force and comparing accessions with the applicant pool. The trends for applicants and accessions without prior service are shown in Figure 6 by AFQT for the years 1988 to 1992. While there are significant declines in applicants after 1990 for all AFQT levels, nevertheless the supply of potential recruits exceeds the stated 1985-1989 quality requirements (Figure 3) at all levels of quality. For example, the total number of male accessions for the active force was approximately 172,000 in 1992. If the 1985-1989 quality requirements were applied to this number, they would generate a need for approximately 105,000 male recruits in Category I-IIIA, 52,000 in Category IIIB, and 15,000 in Category IV that year.

The actual male accessions shown in Figure 6 are quite different from the 1985-1989 stated requirements, being 128,000 for Category I-IIIA, 43,000 for Category IIIB, and only about 300 or so for Category IV. Clearly, the Services are recruiting more high-quality men than called for by these earlier requirements. Moreover, the accession trend lines for fiscal 1990 to 1992 show a striking contrast between high-quality recruits and the other

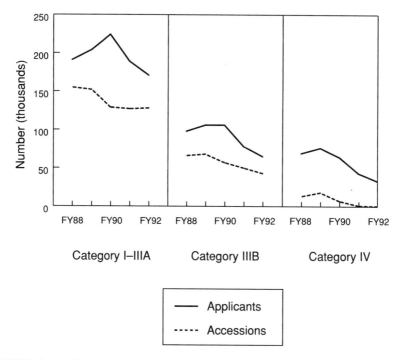

FIGURE 6 Applicants and accessions by AFQT (males, active enlisted force without prior service).

two groups. The number of Category I-IIIA male recruits remained virtually constant at about 128,000 during this period, while the number of Category IIIB fell from 57,000 to 43,000 and the number of Category IV fell from 7,000 to about 300. In other words, virtually all of the reduced accessions were absorbed by Category IIIB and IV recruits.

It would appear from these data that there is indeed an excess of supply at all levels, at least up through fiscal 1992, including that for high-quality categories. The Services have apparently made decisions to take virtually all of the reduction in force from the lower recruit quality levels, particularly Category IV.

There is nothing intrinsically wrong with these recruiting decisions, and indeed the new JPM cost/performance trade-off model may well justify this high-quality mix. When the supply of high-quality recruits exceeds demand, then the marginal cost of recruiting additional high-quality personnel should decline, which makes a high-quality mix more cost-effective (or at least no more costly than one with a lower-quality mix). If this scenario is correct, of course, we would expect to see a corresponding reduction in relative recruiting costs.

To the extent that anyone (such as Congress) questions the very high quality levels between 1990 and 1992, DoD and the Services could offer a more credible defense if the JPM model justified this quality mix. Otherwise, it be would reasonable to assume that the Services are simply following the traditional "more is better" rule for setting manpower quality requirements.

FUTURE TRENDS AND ACTIVE-RESERVE ISSUES

The previous sections have focused primarily on quality requirements for the active forces. This section looks to the future with a focus on the relationship between quality requirements and the active-reserve mix.[6]

The framework we use for this discussion is one of total force demands for enlisted military personnel for both active-duty and reserve forces. Figure 7 displays a simple schematic for total force supply and demand. It highlights the flows into and out of various personnel states without delving into the complexity of cohort timing, which, although important from a personnel management point of view, is not important from our macro perspective. Starting from the left of the chart, we depict the demands for nonprior-service accessions into the active and reserve enlisted ranks from the available pool of male and female youth. Current projections estimate the demands to be about 180,000 for the active force and about 70,000 for the reserve forces.

Active force accessions are enlisted in the active force to serve what is usually called a first term of enlistment. After beginning service, one of several paths may be taken. Attrition during the first term may occur for disciplinary or other performance-related reasons. The two most common paths, however, are either to complete the first term of service and return to civilian life or to reenlist and enter the career force. In the latter stages of this process, successive options to reenlist or return to civilian life occur until mandatory retirement is enforced. The important point about these flows is that those who leave active-duty military service under honorable circumstances are potentially eligible to enter the reserve forces. This is a critically important link between active-duty personnel policy and reserve force structure.

The total force composition of reserve force accessions is also shown in Figure 7. About 70,000 young men and women with no prior military experience are currently planned to be recruited from the available youth population to serve in the reserve forces. A much larger number, about

---

[6]The reserve force includes the Army and Air Force National Guard as well as regular reserve forces in the four Services.

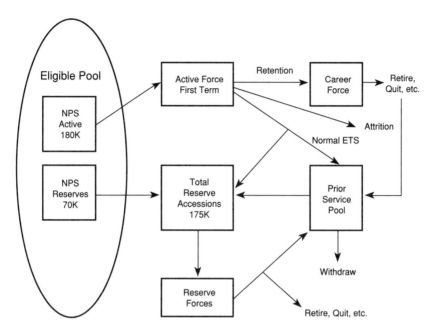

FIGURE 7 Total force supply and demand. (ETS = expiration of term of service)

105,000, are to be recruited from the pool of those who have prior service. We show this in the figure as the flow from the prior service pool into the total accessions pool of 175,000. What we referred to above as a "critically important link" between active and reserve forces personnel policies is clearly delineated in this framework. The reserves rely heavily on the flow out of active-duty status. This is only natural. Since reserve personnel serve so little time, the experience they gain on active duty means that less time must be devoted to training while still maintaining an adequately trained force. Indeed, it is questionable whether, for many technical occupational specialties, it would be possible to use personnel without prior service at all. Using prior-service personnel may not only be cost-effective, but may also be the only way to have certain skills and occupations in the reserve forces at a reasonable cost.

With this framework of flows in mind, let us turn to examine what the future may hold. Figure 8 displays data similar to Figure 4 above, and, not surprisingly, portrays the same conclusions. The figure displays two trends. The first is the historical and projected trend of the number of 18-year-old men. The projection is shown only to the year 2000, but subsequent years immediately past 2000 also show a continued upward trend from the trough experienced in fiscal 1993. The second data series depicts the historical

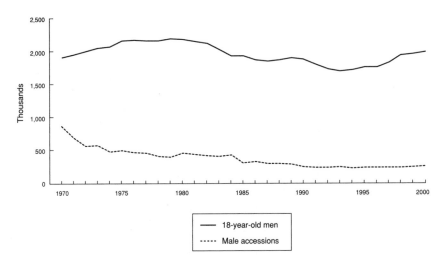

FIGURE 8 Trends in male accessions and available male youth.

and projected trend in male accessions to the active-duty enlisted force.
Obviously, the projected trend is speculative. It assumes that accessions
will be maintained at a level of about 180,000 persons per year, consistent
with the small active-duty force that is currently planned. Despite the
conjectural nature of this forecast, it is clear that raising the number by
40,000 would not alter the fundamental conclusion that we restate here.
Given the increasing youth cohort eligible for military service, it is an
inescapable fact that the gross number of persons projected to be available
for service implies that there should be no shortage of accessions. Furthermore, it also implies that DoD will be able to continue recruiting the numbers of high-quality recruits that it achieves today because of the increasing
pool.
    This conclusion holds for both active and reserve forces. We have
shown the high quality of the active force in previous sections, but the same
high quality is also found in reserve force nonprior-service accessions. Figures 9 and 10 show nonprior-service high school diploma graduate accessions and Category I-III accessions as a percentage of total accessions.
Each component is very high on the two measures of quality and is similar
to the active-duty force. Therefore, there does not appear to be any reason
to think that the available supply of quality and quantity will be deficient in
the near term. This conclusion, that both numbers and quality will not be a
problem in the near future, is buttressed by the fact that the role of women
in the military is increasing significantly. This greatly increases the pool of
high-quality persons available for military service. The availability of more

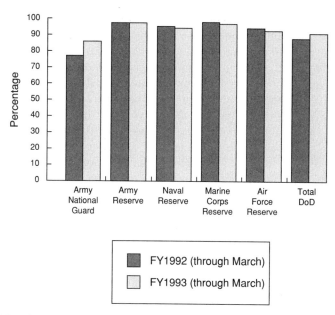

FIGURE 9 High school diploma graduate accessions as a percentage of total accessions.

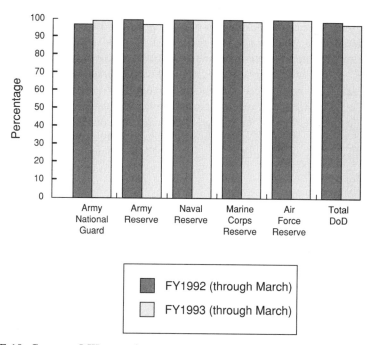

FIGURE 10 Category I-III accessions as a percentage of total accessions.

women for noncombat specialties will increase the number of high-quality men available for combat jobs.

In summary, examining the supply-demand balance until the year 2000 suggests that there appear to be no aggregate shortages of the quantity or quality of nonprior-service accessions. But what about the availability of prior-service personnel for the reserves? The same conclusion pertains, but for a different reason. Because the active-duty force is shrinking, thus increasing the numbers of voluntary and involuntary separations, the available supply of prior-service personnel will be increased for the next several years. Therefore, there will be more than an adequate supply of prior-service personnel available to the reserves over this period. That is not to say that one might not find regional shortages and skill imbalances, but these problems can be managed. So, in general, both the active-duty and the reserve forces should be able to draw from pools that are sufficiently large to ensure adequate numbers of high-quality accessions until the year 2000 or so. Indeed, because the supply of high quality would appear to remain large and thus inexpensive to maintain, there might be room to pursue other goals (e.g., social representation) with the enlisted force.

But lest we appear complacent about the ease with which the future can be managed, let us turn to thinking beyond the turn of the century. Despite the fact that the prior-service eligible pool will remain large for some time, there are some significant management challenges ahead of us.

Consider the maintenance of a cost-effective, all-volunteer, active-duty enlisted force. Because of the issue of costs and productivity, this force tends to be a very experience-intensive force. That is, it is a force with a large proportion of members concentrated in the career force. Consequently, relatively few accessions are needed to support first-term demands, and a very large share of the first-term force transitions into the career force, leaving little left over for the prior-service pool eligible for reserve duty. These force parameters are generally those associated with a "low flow" volunteer force and would be expected of the active-duty force once our military forces stabilize at low levels. If reserve forces do not decline with the active force, then reserve accession demands may become large relative to the size of the prior-service pool. This would inevitably lead to shortages unless significant actions were taken.

There are two very different policy options available to cope with the situation outlined above. One could increase reliance on nonprior-service accessions for the reserves, or one could increase active-duty accessions considerably, making the active-duty force rely much more heavily on first-term personnel and creating much higher flows out to the prior-service enlisted pool available to the reserves.

Both of these options require increased knowledge of, and emphasis on, "quality" accessions. For example, if the reserves rely on more nonprior-service personnel, perhaps quality would have a higher payoff in the reserves than in the active force. Quicker training times would be even more important in the reserves, and better skill retention might also result. The productivity of quality, therefore, might be even higher relative to its cost than in the active force. In a junior force mix, therefore, quality might be the key to a cost-effective mix of personnel. If, however, one adopts a "high flow" volunteer force model to produce more prior-service personnel available for reserve duty, then the cost of quality may rise considerably, seemingly making the All-Volunteer Force more expensive. However, since these associated costs were incurred to pursue reserve strengths, the costs of this policy should be allocated to the reserves, not to the active forces.

As we have discussed, there are no particular near-term concerns regarding quality. Barring serious mismanagement, the active and reserve force structures should be sustainable. However, in thinking about the situation beyond the turn of the century, it is apparent that future active-reserve mix policies will probably conspire to make quality an issue again. Compared with the knowledge we have gained about active-duty enlisted personnel, we know surprisingly little about the relationships between quality and job performance in the reserves. It is not too early to start the research we need to understand these relationships so we can be ready to provide well-informed policy advice when it is certain to be needed.

CONCLUSION

In the preceding pages we have sketched out the changes in forces, force mixes, and demographic trends that have given rise to the various issues surrounding quality. As we said at the outset, the questions today are much different than those that existed in 1980. In the early period, quality was at the very heart of the viability of the All-Volunteer Force. It appeared doubtful that quality standards could even be maintained at low levels in the face of the declining youth cohort and expected compensation levels.

Today the question seems to be one of how much quality is enough. Indeed, some people would say the All-Volunteer Force has too much quality, that the supply of quality exceeds the demands. There is no doubt that this position is overly simplistic, but at the very least we should raise the same type of question that was raised during the quality crisis 15 years ago. At that time, there was consensus that military personnel and recruiting expenditures had to be raised to meet quality requirements. Should we not also ask today, with our unprecedented levels of high quality, whether we are spending more on personnel and recruiting costs than is necessary to maintain a reasonable level of quality?

Finally, despite the fact that we expect the next several years to produce adequate supplies of both quality and quantity to both active and reserve forces, we also raise questions about management of the total force after the current reduction in force has been fully implemented. Beyond the year 2000 we may face significant trade-offs of quality for quantity in the active or reserve forces depending on what management strategies of force mix are pursued. We have argued that we need to increase our knowledge of cost/performance trade-offs in the reserves so as to better position DoD to make sound management choices over the next several years.

Of even higher priority, we have suggested that it is important to use the cost/performance methods developed in the JPM Project to evaluate the current high-quality force mix that is emerging today. If a cost/performance trade-off model could defend the increases in manpower quality and costs that took place after 1980, will the JPM model justify the high levels of manpower quality and recruiting costs today? Application of the JPM model to these questions would put the methodology in the policy mainstream and solidify its value to manpower policy managers in coming years.

REFERENCES

Armor, David J., Fernandez, Richard L., Bers, Kathy, and Schwarzbach, Donna
  1982   *Recruit Aptitudes and Army Job Performance.* R-2874-MRAL. Santa Monica, Calif.: The Rand Corporation.
Cooper, Richard V.L.
  1977   *Military Manpower and the All-Volunteer Force.* R-1450-ARPA. Santa Monica, Calif.: The Rand Corporation.
Fernandez, Richard L., and Garfinkle, Jeffrey B.
  1985   *Setting Enlistment Standards and Matching Recruits to Jobs Using Job Performance Criteria.* R-3067-MIL. Santa Monica, Calif.: The Rand Corporation.
Vineberg, Robert, and Taylor, Elaine N.
  1972   *Performance in Four Army Jobs by Men at Different Aptitude (AFT) Levels: 3. The Relationship of AFQT and Job Experience to Job Performance.* Technical Report 72-22. Alexandria, Va.: Human Resources Research Organization.

Part II:
Job Performance
Measurement Issues

The JPM Project was an ambitious effort to measure on-the-job performance of enlisted military personnel. The project offered researchers the opportunity to test hypotheses about differences between hands-on measures of job performance and paper-and-pencil surrogates of performance, about ethnic group differences in performance, and about gender difference in performance. Among the many concerns of the JPM Project and the committee, three were of primary interest. The first concern centered on the adequacy of test administration activities, such as scheduling, test security, and administration consistency from one individual to the next. A second concern was that the scaling of hands-on performance scores should go beyond rank ordering. That is, there was a need for a score scale that could be interpreted in terms of, at a minimum, acceptable and unacceptable performance, and preferably at finer gradations. A third concern centered on how job tasks should be selected. The committee recommended that stratified, random sampling of tasks be used rather than purposive sampling. They argued that purposive selection might capture only a certain type of task amenable to testing and might not be representative of the job, whereas stratified random sampling provided an unbiased selection of representative tasks and could more easily be defended.

Part II of this volume contains two papers dealing with various aspect of job performance measurement. In the first paper, Lauress Wise addresses the three concerns listed above in a thorough analysis of issues surrounding the validity of the JPM data and data from other sources such as the Synthetic Validity (SYNVAL) project. He also examines the appropriateness of using these data for setting performance goals in the cost/performance trade-off model.

In the second paper in this section, Rodney McCloy pursues further the critical issue of generalizing performance results from jobs on which performance has been measured to jobs for which no data are available. This issue has been of particular importance because only a few jobs were selected for detailed study in the JPM Project and there was a need to generalize the findings to the several hundred jobs performed by first-term enlisted personnel. For the current model, the multilevel regression analysis method was recommended because of its contributions to performance prediction at the job level. The SYNVAL approach was considered but was deemed too time-consuming for the present project. McCloy discusses the application of the multilevel regression analysis in detail.

Setting Performance Goals for the DOD Linkage Model

Lauress L. Wise

The times certainly are changing, particularly for the Department of Defense (DoD). The types of threats to which we must be ready to respond are changing; the size of the forces available to respond to these threats has decreased significantly and is likely to decrease further yet; and the resources available for recruiting, training, and equipping our forces have also declined dramatically. The debate continues as to how much we can afford to spend on defense in the post-cold war era and how much we can afford to cut. Efforts to keep missions, forces, and resources in some kind of balance are now focused on an emerging concept of *readiness*.

The DoD cost/performance trade-off model can play a central role in balancing readiness and resources. Those who sponsored the development of this model could not possibly have anticipated the importance of their efforts, but now that the model is nearing completion, the need for this type of linkage is all too obvious.

The model actually contains two separate linkages. Recruiting resources are linked to the levels of recruit quality, defined in terms of aptitude scores and educational attainment, obtained through application of the recruiting resources. In this first linkage, the model also suggests optimal mixtures of expenditures for recruiters, advertising, and incentives that will yield a given recruit quality mix with the smallest possible total cost. The second linkage is between recruit quality and performance in specific occupational specialties. The full cost/performance trade-off model takes specifications for required performance levels for each different job or family of jobs, determines a recruit quality mix that will yield the desired performance levels, and predicts the recruiting costs required to obtain this mix of recruit quality.

One more linkage is needed. The final step is to tie the emerging concept of readiness to levels of performance in different military specialties. This is, of course, a rather large step. Readiness is not yet a well-defined notion, but it is doubtlessly related to the number and effectiveness of different types of units, with unit effectiveness further related to individual performance levels. The focus of this paper is not, however, on how this last linkage might be achieved, but rather on how we might best set goals for performance levels today, while we are waiting for this final linkage to be created.

The DoD model has allowed us to replace the question of "What level of recruit quality do we need?" with the question "What level of performance do we need?" The goal of this paper is to discuss issues and methods of trying to answer the latter question with information that is currently available to DoD personnel planners and policy makers.

The remainder of this paper is organized into four sections. The first section discusses issues related to the performance metric used in the DoD model. What is the meaning of the performance scale and what is a reasonable answer to the question "What level of performance do we need? The second section describes a normative approach to setting performance level goals. The general idea is to look at predicted performance levels for new recruits at different times and see how these levels varied across time and by job. At the very least, this normative approach will provide plausible ranges for performance level goals. The third section describes criterion-referenced approaches to setting performance level goals. In such an approach, judgments about the acceptability of different levels of performance are analyzed, and then additional judgments about minimum rates of acceptable performance are also collected. The final section lays out suggestions for additional research to further strengthen the support for specific performance level goals.

THE PERFORMANCE METRIC: DESCRIPTION OF THE SCALE FOR HANDS-ON PERFORMANCE QUALIFICATION

Performance in the DoD model is defined as *percent-GO*. The percent-GO scale is derived from the hands-on performance tests developed in the Joint-Service Job Performance Measurement/Enlistment Standards (JPM) Project. A general description of their development is provided by Wigdor and Green (1991:Chapter 4). More detailed descriptions of the development of these measures are provided by the researchers from each Service who worked on their development. Campbell et al. (1990) describe the

Army measures; the Air Force procedures are documented in Lipscomb and Hedge (1988). The description by Carey and Mayberry (1992) of the development and scoring of tests for Marine Corps mechanics specialties is a particularly good source, since this was one of the last efforts undertaken and it built on lessons learned in earlier efforts.

As described by Green and Wigdor (1991), the hands-on performance test scale is an attempt to create a domain-referenced metric in which scores reflect the percentage of relevant tasks that a job incumbent can perform successfully. They were developed as criteria for evaluating the success of selection and classification decisions. An estimate of the proportion of the job that the selected recruit could perform (after training and some specified amount of on-the-job experience and when sufficiently motivated) was judged the most valid measure of success on the job.

In general terms, hands-on test scores do provide at least relative information about success on the job that is reliable and valid. As such they are quite satisfactory for the purposes for which they are intended. There are several issues, however, that affect the level and linearity of the scores derived from them. Among others, these include the sampling of tasks, the scoring of tasks, and the way in which scores were combined across tasks.

Task Sampling

In the JPM Project, a limited number of tasks was selected for measuring performance in each job. If these tasks were selected randomly from an exhaustive list of job tasks, generalization from scores on the hands-on tests to the entire domain of tasks would be simple and easy to defend. This was not, however, the case. Task sampling procedures varied somewhat across the Services. In nearly all cases, there was some attempt to cluster similar tasks and then sample separately from each cluster. In the Army, for example, a universe of up to 700 tasks (or task fragments) was consolidated into a list of 150 to 200 tasks; these tasks were then grouped into 6 to 9 task clusters. One, two, or possibly three tasks were then sampled from each of these clusters. This stratified sampling approach actually leads to a more carefully representative sample of tasks in comparison to simple random sampling. Technically, however, this approach also meant that tasks in different clusters were sampled with different probabilities. Statistical purists might require differential weighting of task results, inversely proportional to sampling probabilities, in order to create precise estimates of scores for the entire domain.

A second and more serious concern with the task sampling procedures is that many types of tasks were either excluded altogether or were selected with very low frequency. There was an attempt to collect judgments about the importance of each task as well as the frequency with which it was performed. Few, if any, low importance or infrequent tasks were selected. In addition, some tasks were ruled out because it would be difficult or dangerous to collect work samples. Tasks for which poor performance by unsupervised recruits could result in damage to individuals or equipment were generally excluded. Tasks that were too easy (did not discriminate among incumbents) and, in a very few cases, too difficult were often excluded from consideration. A consequence of these exclusions is that performance on the sampled tasks generalized more precisely to performance on all job tasks that were judged moderately to highly important, were frequently performed, were not too dangerous, and were challenging enough to be at least a little difficult. This generalization is not necessarily bad, but the relevant domain should be kept in mind when it comes to setting performance standards. Higher performance levels would almost surely be expected for important and frequent tasks than for less important and less frequent tasks, but lower performance levels might be required for less dangerous tasks in comparison to more dangerous tasks; lower performance levels would also be expected for more difficult tasks in comparison to trivially easy tasks. In theory, these differences might offset each other, but to an unknown extent, so that performance on the sampled tasks might not be much different from performance across the entire job domain as called for by the Committee on the Performance of Military Personnel.

Task Scoring

In their "idealized" description of a competency interpretation of the hands-on performance test scores, Green and Wigdor (1991:57) talk about the percentage of tasks in the job that an individual can do. Task performance is not, of course, dichotomous in most cases. For the most part, tasks were divided into a number (from 3 or 4 to as many as 20 or 30) of discrete steps, and criteria were established for successful performance of each of these steps. Naturally there were exceptions: the "type straight copy" task for Army clerks was scored in terms of words-per-minute adjusted for errors; one of the gunnery tasks for Marine Corps infantrymen was scored in terms of number of hits on target. For the most part, however, dichotomous scores were awarded for each of a discrete number of observable steps. In many or most cases, the criterion for successfully performing a step was clear and unambiguous. A mechanic changing a tire either did or did not tighten the lug nuts before replacing the cover, for example. In other cases, the criterion was somewhat arbitrary, as in "the grenade landed within some fixed (but mostly arbitrary) distance of the target" or "the rifle was disassembled within an arbitrarily fixed amount of time." (These standards may have had some strong rationale, but they were not always obvious to the test developers.)

The discrete performance steps varied in terms of their criticality. If a job incumbent had to perform every step successfully in order to be considered successful on the task as a whole, than task success rates would be very low. The scoring generally focused on the process followed more than the overall output. In many cases, it was possible to achieve a satisfactory output even if some of the less critical steps were skipped or poorly performed. Weighting each of the individual steps according to their importance would have required an enormous amount of judgment by subject matter experts and would, in most cases, have led to less reliable overall scores. For purposes of differentiating high and low performers, the percentage of steps performed correctly, without regard to the importance of each step, proved quite satisfactory. When it comes to interpreting the resulting scores, however, it is in most cases impossible to say how many tasks an individual performed correctly because task standards were not generally established. Thus, the real interpretation of the hands-on test scores should be the percentage of task steps that an individual can perform correctly, not the percentage of tasks.

Combining Scores from Different Tasks

For the most part, the scores for each task were put onto a common metric—the percentage rather than number of steps performed successfully—and then averaged to create an overall score. Since the individual task scores were not dichotomous, there was some room for compensation with very high performance on one task (e.g., all steps completed successfully on a difficult task) compensating for somewhat lower performance on another task (e.g., several steps missed on a relatively easier task). As noted above, the tasks were not a simple random sample from a larger domain, and some form of task weighting—either by importance and frequency or by sampling probabilities—would have been possible. The fact that weights were not used should not create problems in interpretation so long as there were not highly significant interactions between task difficulty and importance or frequency. Some bias in the overall scale would also have resulted from the conversion from number to percentage of steps if there were a strong interaction between the number and difficulty of the steps within each task.

Several of the Services also examined different ways of grouping tasks or task steps into clusters in order to create meaningful subscores. The Army analyzed scores from six general task clusters: communications, vehicle operation and maintenance, basic soldiering, identifying targets or threats, technical or job-specific, and safety. Groupings of individual task steps into four knowledge and two skill categories were also analyzed. The Marine Corps created a matrix that mapped task steps onto different "behavioral elements." Although interesting, these subscores did not lead to significant findings and have little bearing on the issue of setting overall performance standards.

In summary, the hands-on test scores derived from the JPM work do lend themselves to an interpretation of job competency. They are scaled in terms of the percentage of steps for important tasks that an individual will perform successfully. It is not unreasonable to interpret these scores in a general sense as the percentage of the central or important parts of the job that the individual can perform successfully. Since a great deal of aggregation is involved in setting performance requirements for the DoD model, a general sense interpretation is probably quite sufficient. Many of the issues raised above could be of significant concern if scores on individuals were being interpreted. Given the imprecision of the prediction of the performance scores from enlistment tests and high school credentials and the highly aggregated nature of the predictions, it seems reasonable to proceed with a general "percentage of job" interpretation.

A NORMATIVE APPROACH

One approach to setting performance level requirements is to ask what levels we have experienced in the past. This is essentially a normative approach in which requirements for future years are tied to norms developed from prior years. If an important objective of the DoD model is to determine whether current quality levels are sufficient or perhaps excessive, then this normative approach is entirely circular, since performance level requirements will be tied back to current quality levels. Furthermore, we would be better off simply using aptitude scores to define quality requirements, since very little new information would be generated in linking performance requirements back to current or past quality levels.

At a more detailed level, however, several interesting questions can be addressed through analyses of normative data. First is the question of the degree of variability in predicted performance levels across jobs. It may well be, for example, that observed differences in recruit quality are evened if high-quality recruits are more likely to be assigned to difficult jobs. A high-quality recruit assigned to a difficult job may end up being able to successfully perform the same percentage of job tasks as a lower-quality recruit in an easier job. If this were the case, then performance level requirements might generalize to new jobs more easily than quality requirements would.

Another question is how much predicted performance levels have varied over time, overall and by job. If performance levels have varied considerably, then using past performance levels to set future requirements would be questionable. If, however, performance levels (and performance level differences among jobs) are relatively stable across time, using past performance levels as a benchmark would be more defensible, although, with dramatic changes in force levels, job requirements may not be as stable in the future as they have been in the past.

Samples

To examine these questions, the fiscal 1982 and 1989 accession cohorts were selected for analysis. The 1982 cohort was the earliest cohort for which the current form of the Armed Services Vocational Aptitude Battery (ASVAB, beginning with forms 8/9/10), was used exclusively in selection. This is important because the performance prediction equations in the linkage model are based on the subtests in the current ASVAB. Earlier ASVAB forms had different subtests and a number of assumptions would be required in generating AFQT and technical composite scores from these prior forms for use in the prediction equation. The 1989 cohort was the most recent for which data on job incumbents with at least two years of service are available. In addition, recruits from this cohort participated extensively in Operation Desert Storm and so some global assessment of their readiness is possible.

For each cohort, the active-duty roster as of 21 months after the end of the enlistment year was examined to identify incumbents in the 24 JPM specialties. The primary military occupational specialty (MOS) at time of enlistment was considered as the basis for sorting recruits into jobs, but it was discovered that many recruits are not enlisted directly into several of the JPM specialties. Consequently, it was decided to select for the JPM specialties on the basis of MOS codes at about 24 months of service. This decision meant that recruits who left service prior to 24 months were not included, and we were thus not modeling the exact enlistment policies. However, examining score distributions among job incumbents considered successful had many advantages and was deemed entirely appropriate.

The JPM samples included 24 different specialties. One of these specialties, Air Force avionics communications specialist, was deleted from the current study. The specialty code was changed prior to the 1989 accession year, and it was not possible to determine whether there was an appropriately comparable specialty.

Variables

ASVAB scores of record were obtained. A small number of cases in the 1982 cohort had enlisted using ASVAB forms 5, 6, or 7. These cases were deleted from the analyses since the ASVAB tests included in the AFQT and technical composites were not all available in these forms. Educational credential status was also obtained and coded as either high school graduate or nongraduate. In these analyses, recruits with alternative high school diplomas were counted among the nongraduates.

Job performance prediction equations for the JPM specialties that were developed in the Linkage Project (McCloy et al., 1992) were used. These equations use AFQT and technical composite scores (expressed as sums of standardized subtest scores), educational level, time in service, and the interaction (product) of time in service and the technical composite in a prediction equation. The weights for each predictor are determined from job analysis information from the *Dictionary of Occupational Titles*. A constant value of 24 months was used for time in service so that predictions would reflect "average" first tour performance for a 3- to 4-year enlistment. Since predicted performance is a linear function of time in service, the average across the first tour will be equal to the value predicted for the midpoint. Ignoring the first six months as mostly training time, the midpoint would occur at month 21 for a 3-year tour and at month 27 for a 4-year tour.

Analyses

The primary results were summarized in an analysis of variance with MOS (23 levels) and accession year (2 levels) treated as independent factors and predicted performance and the AFQT and technical composites each analyzed as dependent variables. One hypothesis tested with these analyses was that predicted performance might show smaller differences among jobs in comparison with the AFQT or technical composites. Separate intercepts (determined from job characteristics) were estimated for each job. It was plausible to believe that required performance score levels might be reasonably constant across jobs, even if input quality was not.

A second hypothesis tested was that predicted performance would show relatively smaller differences across recruiting years in comparison to the AFQT and technical composites, since predicted performance combines both AFQT and technical composite scores and the latter might be less affected by differences in recruiting conditions than the AFQT.

Findings

Across all jobs and both recruiting years, the mean test level was 68.4. Table 1 shows the sample sizes, and the mean and standard deviation of predicted performance scores for each job and entry cohort. Mean predicted performance scores by job and year are also plotted in the table. Table 2 shows the means for the AFQT and technical composites and predicted performance by year. These means are adjusted for differences in the MOS distributions for the two years. Table 3 shows F-ratios testing the

TABLE 1 Mean Predicted Performance by Job and Entry Cohort

| Service | MOS | 1982 | | | 1989 | | | Mean |
		N	Mean	S.D.	N	Mean	S.D.	89–82
Army	11B	10657	62.49	3.42	10341	62.71	3.15	0.22
	13B	5178	60.33	3.74	4065	59.69	3.55	−0.64
	19E	2585	64.98	3.27	2892	64.83	3.34	−0.15
	31C	1497	71.35	2.94	1615	73.04	1.88	1.69
	63B	1915	74.17	3.00	3123	73.84	2.91	−0.33
	64C	3079	59.76	2.85	3902	59.45	3.06	−0.31
	71L	3455	63.98	3.45	1490	64.72	2.26	0.74
	91A	2863	66.10	2.71	4013	65.69	2.83	−0.41
	95B	3397	69.56	2.12	3725	68.85	2.23	−0.71
USAF	122	205	67.14	2.95	235	67.97	2.15	0.83
	272	587	69.11	2.27	623	69.77	1.83	0.66
	324	355	76.17	1.89	152	75.52	2.09	−0.65
	328	432	78.04	1.73	0			
	423	1111	68.32	2.48	711	69.52	2.03	1.20
	426	867	78.60	2.53	744	78.87	2.12	0.27
	492	210	71.73	3.00	146	71.86	1.94	0.13
	732	816	62.08	2.87	765	63.05	2.07	0.97
Navy	ET	1990	74.85	2.04	1547	74.43	2.17	−0.42
	MM	3133	75.78	3.08	3805	75.30	3.07	−0.48
	RM	2202	70.09	3.07	2075	69.54	2.54	−0.55
USMC	031	4392	61.39	3.21	3871	61.23	3.04	−0.16
	033	947	61.98	3.01	1026	61.07	2.74	−0.91
	034	960	78.56	4.87	1084	77.29	4.68	−1.27
	035	1001	65.15	2.92	1421	65.71	2.58	0.56
	Average	53834	66.13		53371	66.26		0.12

significance of differences across years, MOS, and the year-by-MOS inter-action for these same three variables.

The first significant finding from these analyses was that there was virtually no change in mean predicted performance between the 1982 and the 1989 cohorts, overall or for any of the jobs analyzed. A statistically significant mean gain in AFQT was offset by a significant mean drop in technical scores between the 1982 and 1989 cohorts, resulting in no signifi-cant difference in predicted performance. Second, there was some consis-tent variation among jobs in predicted performance levels, with lows of around 60 for Army field artillery (13B) and truck driver (64C), Air Force 732 and two of the Marine Corps infantry jobs and highs above 75 for Air Force 328 and 426 jobs and Marine Corps 034. This variation is consistent with the assumption that higher competency levels might be required in more critical or complex jobs. The variation across jobs in predicted per-formance was much more significant (much greater F-ratio) than the varia-

TABLE 2 Overall Mean Aptitude and Predicted
Performance Scores for Fiscal 1982 and Fiscal 1989
(Adjusting for MOS Differences)

Fiscal Year	AFQT (Sum of SS)	Technical Composite	Predicted Performance
1982	207.52	162.08	68.42
1989	213.50	158.30	68.43
Average	209.51	160.19	68.43

Note: The means were estimated main effect means from an analysis of
variance and are adjusted for differences in the numbers of individuals in
each MOS across the two years. In these analyses, unweighted averages of
the MOS means are used resulting in slightly different values than the results
in Table 1 where each MOS average was weighted by the number of accessions in the indicated year.

SS = standardized subtest scores.

TABLE 3 F-Ratios Testing Components of Variance for Aptitude and
Predicted Performance Scores (Based on 106,663 observations)

Component	AFQT (Sum of SS)	Technical Composite	Predicted Performance
Year (df = 1)	723.0	341.2	0.2
MOS (df = 22)	1074.4	905.4	15,628.1
Year * MOS (df = 22)	58.4	38.01	43.0

SS = standardized subtest scores.
df = degrees of freedom.

tion in AFQT and technical scores. One reason for this is that the within-
job variance of predicted performance is small in comparison to the within-
job variance of the predictor composites. Predicted performance is critical
for all jobs, and so is restricted in range. The predictor composites, particularly the technical composite, are not as critical for all jobs. Each composite is less restricted in range for those jobs for which it is less critical,
leading to greater *average* within job variation. A conclusion that follows
from this finding is that it is probably not sufficient to use a single average
performance level for all jobs. Consequently, some judgmental procedure
is needed to capture essential differences in performance level requirements
for different jobs.

CRITERION-REFERENCED APPROACHES

The Army investigated alternative approaches for setting job performance standards as part of the Synthetic Validity (SYNVAL) Project (Wise et al., 1991; Peterson et al., 1990). This project had two major objectives: (1) to investigate ways of generalizing performance prediction equations from a sample of jobs for which criterion data were available to the entire population of jobs and (2) to investigate ways of setting performance level standards for different jobs. To the extent that performance level standards could be linked to the test scales, the second objective relates directly to the topic of this chapter.

The SYNVAL project was conducted in three phases. The first phase involved pilot tests of job description and standard setting instruments for three jobs. The second phase involved a larger data collection with revised instruments and procedures on a larger sample of 7 jobs. In the final phase, further revisions to instruments and procedures were administered for a sample of 12 jobs to test additional issues, including generalization to one job for which no criterion data were available. The most directly relevant results with respect to standard setting come from Phases II and III.

Most work on standard setting has involved identification of a minimal passing score on a certification or criterion-referenced examination. This might seem to be exactly what is needed in setting performance standards for use with the DoD model. A common concern in education with *minimum* competency examinations is that they provide little motivation for students to achieve at levels well above the minimum. In setting enlistment standards, it is reasonable to ask whether it is acceptable to have all enlistees at the same minimum level or whether it would not be better to have a mix of skill levels within each occupational specialty. Particularly in situations involving teamwork, a mix of skill levels may be more optimal than inordinate homogeneity of skill levels. In the SYNVAL project, four different skill levels were defined for each job. These skill levels were tied to operational decisions that supervisors would make about job incumbents in an effort to derive cost implications for the different performance levels:

Unacceptable: the recruit cannot perform the job, is not likely to become an acceptable performer with additional training, and should be discharged;

Marginal: the recruit is not performing acceptably and should be given additional training to bring performance up to standard;

Acceptable: the recruit is performing at an acceptable level and making a positive contribution to force readiness; and

Outstanding: the recruit is performing well above minimal standards and should be given a promotion or other recognition for superior performance.

In setting performance level goals for the DoD model, standards for minimum performance should not be confused with performance level goals. Performance goals for the model should reflect a desired mix of abilities above the minimum. Ideally, economic analyses would be used to identify the mix of marginal, acceptable, and outstanding performers that is most cost-efficient for each job. Overall performance goals would then reflect an average of scores from these three performance levels, with each level weighted according to this optimal mix. As indicated below, the SYNVAL project results speak primarily to the first step of defining the different performance levels. Examples of the distributions across performance levels for incumbents in different jobs are provided, but economic analyses to identify more optimal mixes remain to be done.

Phase II Design

Five different standard-setting instruments were administered to subject matter experts in the Phase II sample of jobs as described in Peterson et al. (1990). Standards obtained from the different protocols were compared with each other. The level of agreement among judges on the standards they provided was also examined for each approach. It was important that there be adequate agreement among judges on the standards before the standards can be used to determine selection criteria.

The overall objective of this project was to generalize to jobs for which no performance data were available. For this purpose, neither standards for individual tasks nor standards for the job as a whole were judged useful, although, in retrospect, the data on overall job standards were quite informative. The primary focus was on performance dimensions defined in terms of families of related job tasks. The job performance dimensions used for the standard-setting exercises came from a preliminary version of the hybrid taxonomy (used in the job analysis; see Peterson et al., 1990). A preliminary set of 24 dimensions were identified based on job components contained in the task categories and job activities taxonomy. Not all 24 dimensions were applicable for the Phase II jobs and thus the summary tables show only those dimensions relevant to Phase II jobs.

Three different proficiency categories based on the three minimum performance levels (cutoffs) that defined the performance levels described were examined. The three proficiency categories were unacceptable (less than marginal), unacceptable and marginal combined (less than acceptable), and outstanding (greater than acceptable). The last category was described as outstanding rather than less than outstanding to enhance interpretability.

Three different standard-setting protocols are referred to here:

1. *Soldier-Based Protocol (Soldier Method)*. Under this protocol, judges

were asked to estimate the percentage of current job incumbents who are performing at each of the four levels of acceptability (e.g., what percentage is unacceptable) on a given performance dimension. This approach assumes that empirical data on soldier performance are available (in the form of hands-on tests scored GO/NO-GO) on a representative sample of the soldiers in question so that these "percentage-performing" estimates can be related to actual performance scores.

2. *Critical Incident Protocol (Incident Method).* Under this protocol, judges were presented with incidents that reflected varying levels of effectiveness on a particular performance dimension and asked to judge, for each incident, the acceptability level of soldiers whose typical performance was described by the incident.

3. *Task-Based Protocols (Task-Hypothetical Soldier, Task-Detailed Percent-GO, and Task-Abbreviated Percent-GO Methods).* Under these protocols, judges were presented with a list of specific tasks within each performance dimension (possibly from different MOS) and asked to make judgments about minimum percent-GO scores that a soldier should achieve to qualify as marginal, acceptable, and outstanding performers. Three types of judgments were collected. In the first condition, the hypothetic soldier (HS) approach, judges were presented detailed sets of hands-on test score sheets and corresponding summary percent-GO scores for 10 hypothetical soldiers and asked to rate the acceptability of each of these hypothetical soldiers (Task-HS Method). In the second condition, the detailed percent-GO (DPG) approach, judges were asked to rate the minimum percent-GO score for each level of acceptability on each specific task used to illustrate the dimension (Task-DPG Method). In the third condition, the abbreviated percent-GO approach, judges were given a list of tasks without detailed percent-GO scores or actual score sheet examples and asked to rate minimum percent-GO scores for tests on these types of tasks (Task-APG Method).

The five different standard-setting methods involved judgments that used very different metrics. The soldier-based method asked about the percentage of soldiers performing at each acceptability level; the critical incident method used a series of discrete behavioral items; and the task-based methods used judgments about acceptable levels of percent-GO scores.

A critical question in this research was the extent to which the different methods led to similar or distinct ability requirements. To answer this question, it was necessary to convert the standards derived from each approach to a common metric, making it possible to determine whether one of the methods led to significantly stricter or more lenient standards than the others and also to compare the level of agreement among judges using this same metric.

The soldier-based metric (percentage of soldiers performing at each level) was used as the basis for comparison. If standards set with the other methods led to very different assessments of the percentage of soldiers performing at each level (in comparison to the judges' direct assessment), then the validity of these other methods would be questionable. Data from the Army's Project A on samples of incumbents in each MOS were used to estimate the percentage of soldiers performing above or below each of the standards set. The specific methods used to estimate the percentages of soldiers performing at or below specific critical incident or Percent-GO score levels are detailed in Whetzel and Wise (1990).

Phase II Results

Table 4 shows the means and standard deviations of the judges' ratings of the percentage of soldiers performing at each acceptability level for each combination of performance dimension and MOS. There are some distinct differences in the judges' estimates of soldiers' ability across different MOS and dimensions. For example, for MANPADS crew members (MOS 16S), soldiers had high acceptability ratings for performance dimension 7 (detect targets), but relatively lower acceptability ratings on dimension 15 (operate vehicles). These differences reflect, in part, the appropriateness or importance of the dimension for the MOS (e.g., all crew members detect targets, but not all have to operate vehicles).

The standard deviations in the table are a measure of the degree of agreement among judges. These numbers also give an indication of the potential appropriateness of the dimension for the MOS. When there is more significant disagreement among judges, it may be because the dimension is poorly described or is not clearly appropriate for the MOS in question. To a certain extent, the standard deviations are related to the means— when there is more disagreement, the means tend to be closer to 50 percent of soldiers performing at a particular proficiency level. (Only very high or low scores are possible if nearly all of the judges consistently give high or low ratings.) In some cases, however, the standard deviations are greater than the means (e.g., the percentage of 16S soldiers rated unacceptable on the task operate vehicles or the percentage of motor transport drivers, MOS 88M, rated unacceptable on the task navigate). This can happen only when the distribution of ratings is highly skewed, with most judges giving low ratings (hence a low mean) and a few judges giving very high ratings (leading to a large standard deviation).

Similarly detailed analyses of results from each of the other methods are reported in Whetzel and Wise (1990). Table 5 shows comparisons of the overall results from each of the five methods. The methods varied considerably in terms of "leniency": the soldier method suggested only 15 percentage of current job incumbents performed unacceptably and 25 per-

TABLE 4 Mean and Standard Deviation of the Percentage of Soldiers at Each Level, by Dimension and MOS: Soldier Method

Level	16S	19K	67N	76Y	88M	91A	94B	Avg.
Performance Dimension	Mn/SD	Mn/SD	Mn/SD	Mn/SD	Mn/SD	Mn/SD	Mn/SD	Mn/SD
Percentage Unacceptable								
2. Crew Served Wpns.	14/18	08/07						11/13
3. Tactical Mvmnts.	12/12	10/09						11/11
4. Navigate					21/22			21/22
5. First Aid						16/18		16/18
7. Detect Targets	8/07	10/09	10/14					09/10
8. Repair Mech. Sys	16/14	10/08	12/10		15/12			13/11
10. Use Tech Refs.				20/19				20/19
11. Pack and Load	17/19	07/10		19/19	14/14		14/10	14/14
13. Operate/Install			11/13				15/15	13/14
15. Operate Vehicles	21/26	05/07			8/06			11/13
16. Type				25/24				25/24
17. Record Keeping			17/15	16/17	20/19	19/19		14/18
18. Oral Comm.	16/14	12/12				13/12		14/13
19. Written Comm.				27/25		15/12		21/19
22. Medical Treatmnt						11/11		11/11
23. Food Preparation							13/12	13/12
24. Leadership						16/15		16/15
Average	15/16	09/09	13/13	21/21	16/15	15/15	14/12	15/12
Sample Size	563	378	162	235	250	342	129	1,807

TABLE 4 (Continued)

Percentage Unacceptable
Percentage Less Than Acceptable

Level	16S	19K	67N	76Y	88M	91A	94B	Avg.
Performance Dimension	Mn/SD	Mn/SD	Mn/SD	Mn/SD	Mn/SD	Mn/SD	Mn/SD	Mn/SD
2. Crew Served Wpns.	32/20		23/16					28/18
3. Tactical Mvmnts.	33/16		27/21					30/19
4. Navigate					40/23	37/26		40/23
5. First Aid								37/26
7. Detect Targets	24/15	28/21	26/18					26/18
8. Repair Mech. Sys.	42/21	30/19	26/13		37/20			34/18
10. Use Tech Refs.				40/22				40/22
11. Pack and Load	38/25	25/23		42/23	33/21		11/07	30/20
13. Operate/Install			27/20				10/07	18/14
15. Operate Vehicles	36/23	18/16			24/15			26/15
16. Type				47/27				47/27
17. Record Keeping			37/22	39/24	42/23	26/11		36/20
18. Oral Comm.	36/20	31/24				36/23		34/22
19. Written Comm.				52/26		40/21		46/24
22. Medical Treatmnt.						30/21		30/21
23. Food Preparation							13/17	13/17
24. Leadership						43/23		43/23
Average	34/20	26/20	29/18	44/24	35/20	35/21	11/10	25/19
Sample Size	563	378	162	235	250	342	129	1,807

Percentage Outstanding

2. Crew Served Wpns.	14/16	11/09						13/13
3. Tactical Mvmnts.	17/18	08/10						13/14
4. Navigate					13/12			13/12
5. First Aid						12/14		12/14
7. Detect Targets	26/24	10/08	08/05					15/13
8. Repair Mech. Sys.	14/17	08/08	12/11		15/18			12/14
10. Use Tech Refs.				14/17				14/17
11. Pack and Load	15/17	09/11		16/20	14/11		11/07	13/13
13. Operate/Install			13/16				10/07	12/12
15. Operate Vehicles	18/20	10/11			19/19			16/17
16. Type				15/18				15/18
17. Record Keeping			08/08	20/23	11/07	11/13		13/13
18. Oral Comm.	17/19	09/11				13/14		13/15
19. Written Comm.				16/19		12/16		14/18
22. Medical Treatmnt.						12/14		12/14
23. Food Preparation							13/17	13/17
24. Leadership						13/17		13/17
Average	17/19	09/10	10/10	16/19	14/13	12/15	11/10	11/14
Sample Size	563	378	162	235	250	342	129	1,807

Note: A total of 24 performance dimensions were available for standard setting. However, not all were relevant for Phase II MOS. Only the relevant dimensions are shown in the Tables. Column averages are unweighted averages of means and standard deviations for the different MOS. 16S = MANPADS Crewmember, 19K = Armor Crewman, 67N = Utility Helicopter Repairer, 76Y = Unit Supply Specialist, 88M = Motor Transport Operator, 91A = Medical Specialist, 94B = Food Service Specialist.

TABLE 5   Summary of Phase II Rating Results by Judgment Method

Level/Method	Mean	SD	Reliability[a]
Percentage Unacceptable			
Soldier Method	15	12	.07
Incident Method	22	17	.18
Task-HS Method	34	24	.15
Task-DPG Method	39	21	.29
Task-APG Method	43	22	.42
Percentage Unacceptable or Marginal			
Soldier Method	25	19	.09
Incident Method	29	20	.20
Task-HS Method	58	25	.12
Task-DPG Method	63	21	.28
Task-APG Method	63	22	.40
Percentage Outstanding			
Soldier Method	15	17	.12
Incident Method	18	18	.11
Task-HS Method	9	11	.13
Task-DPG Method	10	12	.18
Task-APG Method	11	13	.23

[a]The reliability for each performance level is estimated as the ratio of true variation in the percentage of soldiers across MOS and performance dimensions to the total variation, including differences among judges. These reliabilities apply to individual judgments; the reliabilities of means across several judges can be estimated using the Spearman-Brown formula: $r_n = n * r_1 / (1 + (n - 1) * r_1)$, where r_1 is the single rater reliability and n is the number of judges.

centage were less than fully acceptable, while the task-APG method implied that 43 percent were at unacceptable and 63 percent were at less than fully acceptable levels.

There were also notable differences in the reliabilities associated with the different methods. The task-based methods, particularly those based on percent-GO score ratings, had significantly higher single-rater reliabilities than the other methods. This appears to be a result of stereotypical beliefs that 60 or 70 percent correct should be the minimum "passing" score.

Comparison of Task-DPG and Task-APG Results

The task-DPG and task-APG methods are of particular interest because they use the same Percent-GO scale used in the DoD model.   The only

TABLE 6 Comparison of Minimum Percent-GO Scores by Acceptability Level, for the Task-Based Detailed and Abbreviated Percent-GO Methods

	Detailed Percent-GO		Abbreviated Percent-GO	
Category	Mean	SD	Mean	SD
Marginal	66	12	69	10
Acceptable	78	09	80	08
Outstanding	92	06	93	06

difference between these two approaches is that, for the task-DPG method, a great deal of information is provided about the particular steps (items) that are considered in computing the percent-GO scores. It is reasonable to ask whether this additional information led to different standards or different levels of agreement among judges. In other words, did the extra information help judges to reach a common understanding or just confuse them?

Table 6 shows the means and standard deviations of the percent-GO scores that resulted from each method, rater group, and acceptability level. As can be seen from this table, the APG method usually led to slightly harsher ratings, but also very slightly smaller standard deviations than the DPG method. The differences were minimal at most.

Additional Results from Phase III

In Phase III of the Army Synthetic Validity Project, standard-setting instruments were revised and used to collect data on 12 additional jobs. The task dimensions for which standards were set were also revised to make the dimensions parallel to the major categories in the revised job description instrument.

The task-based standard-setting instrument is most relevant to the issues in this paper. The Phase III version was simplified by eliminating detailed information about the task tests and eliminating the requirements for setting standards for individual tasks. For each performance dimension, three illustrative tasks were listed and then the number of soldiers at or below each percent-GO score level (in increments of 5 from 10 to 100) was provided. Raters were asked to draw lines between the score levels to indicate divisions between different performance levels (unacceptable versus marginal, marginal versus acceptable, and acceptable versus outstanding).

Table 7 shows the mean and standard deviation of the percent-GO cutoffs for each Phase III job, performance dimension, and performance level. As shown in this table, there was reasonable consistency across jobs, with

TABLE 7   Percent-GO Cutoffs for Phase III Jobs Using Revised Task-Based Standard-Setting Instruments

		Minimum Percent-GO for Performance Level:		
MOS	Sample Size	Marginal	Acceptable	Outstanding
12B	76	65.4	79.4	94.6
13B	67	70.4	82.2	94.6
27E	22	63.9	77.9	93.7
29E	28	66.1	80.0	94.7
31C	75	69.1	82.3	96.1
31D	16	62.5	77.6	93.0
51B	75	65.3	79.5	94.7
54B	17	63.4	78.5	95.4
55B	44	62.9	77.8	93.8
95B	36	67.6	81.4	94.9
96B	42	59.1	78.0	94.4
Overall	498	65.1	79.5	94.5

Note: 12B = Combat Engineer, 13B = Cannon Crewman, 27E = TOW/Dragon Repairer, 29E = Radio Repairer, 31C = Single-Channel Radio Operator, 31D = Mobile Subscriber Equipment Transmission System Operator, 51D = Carpentry and Mason Specialist, 54B = Chemical Operations Specialist, 55B = Ammunition Specialist, 95B = Military Police, 96B = Intelligence Analyst.

minimum scores of about 65 percent, 80 percent, and 95 percent for marginal, acceptable, and outstanding levels, respectively. It is difficult to tell the extent to which the small differences among jobs in the cutoff scores are reliable. Different task dimensions and different groups of judges were used with the different jobs, and the variation in results may well be associated with random and systematic factors associated with these differences. In any event, given all of the limitations on the accuracy of domain-referenced interpretations of the percent-GO scales, these differences in cutoffs would not appear to be of practical significance.

Table 8 shows the estimated percentage of current job incumbents at the lower and higher performance levels. The Phase III approach attempted to combine the task-based and soldier-based approaches by providing both criterion information (about the tasks) and normative information (about the proportion of soldiers at each level). As shown in the table, the results reflected this compromise with the proportion of soldiers judged unacceptable (28 percent) or less than fully acceptable (48 percent) falling midway between the Phase II results for the separate soldier and task-based methods (15 to 43 percent and 25 to 63 percent, respectively). The percentage of soldiers at the outstanding level (20 percent) also fell between the extremes of the Phase II methods (12 to 23 percent). Variation in the performance distributions across jobs was somewhat greater in comparison to the variation in score cutoffs, particularly at the high end of the scale. The percent-

TABLE 8 Percentage of Soldiers at Each Performance
Level, Using Revised Task-Based Standard Setting
Instruments

| MOS | Percentage of Job Incumbents Who Are: | | |
	Less Than Unacceptable	Acceptable	Outstanding
12B	27.6	50.4	16.5
13B	32.2	49.9	20.1
27E	22.9	42.4	22.7
29E	26.3	44.6	33.3
31C	31.1	49.1	16.7
31D	24.6	41.1	27.3
51B	27.4	50.5	17.2
54B	25.1	48.6	15.4
55B	27.5	50.5	17.3
95B	35.1	55.3	16.4
96B	26.6	49.4	17.3
Overall	27.9	48.4	20.0

age performing at an outstanding level varied from 15 percent for 54B to 33
percent for 29E.

Summary of SYNVAL Standard-Setting Results

The SYNVAL Project demonstrated both the promise and the difficul-
ties of efforts to define comparable performance categories across different
jobs. There were a number of arguments and concerns about differences in
procedures and instruments and the reliability of individual judgments was
not extremely high. One persistent finding was that standards set using the
hands-on performance tests appeared harsh in comparison with direct estimates of performance level distributions. The consequence, in Phase III,
that over a quarter of all recruits are performing unacceptably and should be
terminated may be difficult to accept. That as many as half of the incumbents would benefit from additional training is much more credible and is
consistent with current refresher training programs. At the upper end, the
definition of outstanding performance is somewhat more subjective, and 20
percent outstanding is not unreasonable.

An important finding from the SYNVAL project was that cutoff scores
for the percent-GO scale on the hands-on performance tests were reasonably similar across jobs. It would be reasonable to adopt 65, 80, and 95
percent cutoffs for all jobs, eliminating a requirement to collect new judgments for each new jobs. This is particularly important since this is the
same metric used in the model. What remains is to identify factors associ-

ated with differences in optimal mixes of performance levels across different jobs. The key question seems to be "When are higher proportions of outstanding performers required?"

CONCLUSIONS

Summary of Findings

The performance metric used with the DoD model was constructed in such a way that a domain-referenced interpretation is at least plausible. There were some potentially offsetting restrictions on the domain of job performance covered by the hands-on performance tests. Although no absolute definition of successful performance was developed at the task level, there were clear criteria for success on individual task steps. For the global purposes required by the model, it is not unreasonable to interpret the scores as the percentage of the job that a recruit can perform successfully.

Normative data for two different entry cohorts showed significant variation across jobs in mean predicted performance but remarkable stability over time periods. Mean predicted performance scores ranged between 60 and 80 percent across different jobs with an overall average of 66 percent. Based on these data, it appears reasonable to use past predicted performance levels in setting performance targets for each job, but generalization across jobs will be somewhat limited.

In a more criterion-referenced approach, the Army Synthetic Validity Project analyzed procedures for defining different levels of job performance that are tied to possible economic consequences associated with good or poor performance. Task-based methods tied to the hands-on performance tests tended to yield stricter standards in comparison to direct judgments about the proportion of soldiers at each performance level. Standards set using the task-based methods were reasonably consistent across jobs. Performance below 65 percent was considered unacceptable, with the implication that the recruit should be discharged; from 65 to 80 percent was considered marginal, with the implication that additional training should be provided; from 80 to 95 percent was considered acceptable; and above 95 percent was considered outstanding with promotion or some other recognition deemed appropriate. Some variation among jobs in the proportion of incumbents at each performance level was observed for each of the standard-setting methods.

The normative and criterion-referenced approaches agreed that there was significant variation across jobs in performance levels. Normative data suggested that performance level targets of about 66 percent were consistent with current accession and readiness levels. The criterion-referenced approach implied that this was a minimally acceptable level and not necessarily a good target for average performance.

Implications for Further Research

If we accept the results of the normative and criterion-referenced approaches summarized above, we can consider setting performance targets for the DoD model by multiplying the targeted number of recruits by an average performance level between 60 and 80 on the hands-on performance test percent-GO scale. Further research would be useful in defending more precise mean predicted performance level targets and, in particular, in supporting differences among occupational specialties in performance level targets. Several specific topics for additional research are discussed below.

Enhancing Performance Level Descriptions

The SYNVAL project attempted to link performance levels to operational decisions about individuals. This linkage was based entirely on expert judgments. A fruitful area for further research would be the development of better descriptions of what individuals at different performance levels can and cannot do. There has been a considerable effort in recent years to establish overall standards for educational achievement for use in interpreting results from the National Assessment of Educational Progress. Part of this process has involved analyses of items answered successfully by students at one level by not at the next lower level. A similar effort with the hands-on test performance levels would help in the development of explicit rationales for economic consequences of performance at specific levels.

Linking Job Characteristics to Performance Distribution Targets

While common performance level descriptions appeared feasible, there was considerable variation across jobs in the proportion of incumbents at each level. More systematic research is needed on the relationship of job characteristics (e.g., task complexity, the extent of teamwork, indicators of criticality of tasks, specific consequences of unsuccessful task performance) to different performance distribution targets. Most particularly, differences among jobs in the need for outstanding performers should be modeled.

Linking Performance Distribution Targets to
Unit Effectiveness and Readiness

As more concrete conceptions of factors relating to readiness emerge, it would be useful to relate these factors to level and heterogeneity of the performance of individuals in different units. In particular, analysis data from unit training exercises should prove useful in linking individual performance levels to indicators of unit effectiveness.

REFERENCES

Campbell, C.H., Ford, P., Rumsey, M.G., Pulakos, E.D., Borman, W.C., Felker, D.B., de Vera, M.V., and Riegelhaupt, B.J.
1990 Development of multiple job performance measures in a representative sample of jobs. *Personnel Psychology* 43:277-300.

Carey, N.B., and Mayberry, P.W.
1992 Development and scoring of hands-on performance tests for mechanical maintenance specialties. *CNA Research Memorandum 91-242*. Alexandria, Va.: Center for Naval Analyses.

Eitelberg, M.J.
1988 *Manpower for Military Occupations.* Alexandria, Va.: Office of the Assistant Secretary of Defense (Force Management and Personnel).

Green, B.F, Jr., and Wigdor, A.K.
1991 Measuring job competency. In A.K. Wigdor and B.F. Green, eds., *Performance Assessment for the Workplace, Volume 2.* Committee on the Performance of Military Personnel. Washington, D.C.: National Academy Press.

Laurence, J.H., and Ramsberger, P.F.
1991 *Low-Aptitude Men in the Military: Who Profits, Who Pays?* New York: Praeger Publishers.

Lipscomb, M.S., and Hedge, J.W.
1988 *Job Performance Measurement: Topics in the Performance Measurement of Air Force Enlisted Personnel.* Technical Report AFHRL-TP-87-58. Brooks Air Force Base, Tex.: Air Force Human Resources Laboratory.

McCloy, R.A., Harris, D.A., Barnes, J.D., Hogan, P.F., Smith, D.A., Clifton, D., and Sola, M.
1992 *Accession Quality, Job Performance, and Cost: A Cost-Performance Tradeoff Model.* HumRRo Report No. FR-PRD-92-11. Alexandria, Va.: Human Resources Research Organization.

Peterson, N.G., Owens-Kurtz, C., Hoffman, R.G., Arabian, J.M., and Whetzel, D.L.
1990 *Army Synthetic Validity Project: Report of Phase II Results. Volume I.* Army Research Institute Technical Report 892. Alexandria, Va.: United States Army Research Institute for the Behavioral and Social Sciences.

Whetzel, D.L., and Wise, L.L.
1990 Analysis of the standard setting data. In N.G. Peterson, C. Owens-Kurtz, R.G. Hoffman, J.M. Arabian, and D.L. Whetzel, eds., *Army Synthetic Validity Project: Report of Phase II Results, Volume I.* Army Research Institute Technical Report 892. Alexandria, Va.: United States Army Research Institute for the Behavioral and Social Sciences.

Wigdor, A.K., and Green, B.F., Jr., eds.
1991 *Performance Assessment for the Workplace, Volume 1.* Committee on the Performance of Military Personnel. Washington, D.C.: National Academy Press.

Wise, L.L., Peterson, N.G., Hoffman, R.G., Campbell, J.P., and Arabian, J.M.
1991 *The Army Synthetic Validity Project: Report of Phase III Results, Volume I.* Army Research Institute Technical Report 922. Alexandria, Va.: United States Army Research Institute for the Behavioral and Social Sciences.

Predicting Job Performance Scores Without Performance Data

Rodney A. McCloy

Military manpower and personnel policy planners have pursued the goal of documenting the relationship between enlistment standards and job performance for over 10 years (Steadman, 1981; Waters et al., 1987). Prior to the Joint-Service Job Performance Measurement/Enlistment Standards (JPM) Project begun by the Department of Defense (DoD) in 1980, proponents of validity studies examining the Services' selection test, the Armed Services Vocational Aptitude Battery (ASVAB), had primarily relied on measures of success in training as criteria. The catalysts for the enormous JPM effort included the misnorming of ASVAB forms 6 and 7 that resulted in the accession of a disproportionate number of low-aptitude service men and women, a decrease in the number of 18-21-year-olds (i.e., the enlistment-age population), and the perpetual requirement of high-quality accessions (Laurence and Ramsberger, 1991). These events simultaneously focused attention on the need to relate the ASVAB to measures of job performance and the absence of such measures.

The outcome of this series of events was an all-Service effort to measure job performance and to determine the relationship between job performance and military enlistment standards. The steering committee for this effort established general guidelines for the work but encouraged diverse approaches to performance measurement in the interests of comparative research. To this end, each Service conducted its own JPM research program with its own specific goals and questions. As a result, the measures and samples across the Services are sometimes quite different. For example, the Army wished to limit the effect that job experience would have on the results from their JPM research project (Project A; Campbell, 1986). Hence, the range of months in service in the Army sample is small.[1] In contrast, the Marine Corps was keenly interested in the effect of job experience on performance and developed its performance measures to be applicable to soldiers in both their first and second tours. Accordingly, the range of experience in the Marine Corps sample is relatively large.

Although each Service developed several job performance measures (Knapp and Campbell, 1993), the Joint-Service steering committee selected the work sample (or "hands-on") performance test as the measure to be "given [resource and scientific] primacy" and to serve as "the benchmark measure, the standard against which other, less faithful representations of job performance would be judged" (Wigdor and Green, 1991:60). Although some have questioned whether hands-on measures are the quintessential performance measures (Campbell et al., 1992), there is little debate over the notion that hands-on measures provide the best available assessment of an individual's job proficiency—the degree to which one *can* perform (as opposed to *will* perform) the requisite job tasks.

The advantages and disadvantages of hands-on measures are well known. The scientific primacy given them by the Committee on the Performance of Military Personnel is justifiable at least in part by their face validity and their being excellent measures of an individual's task proficiency. However, there can be limitations regarding the types of tasks they can assess (e.g., it would be difficult to assess a military policeman's proficiency at riot control using a hands-on measure), and their resource primacy is virtually required given their expense to develop (see Knapp and Campbell, 1993, and Wigdor and Green, 1991, for detailed descriptions of the development of the hands-on performance tests). To highlight this point, consider that hands-on tests were developed for only 33 jobs as part of the JPM Project (Knapp and Campbell, 1993), the most extensive performance measurement effort ever conducted.[2]

Much has been gained from the JPM research. The ASVAB has been shown to be a valid predictor of performance on the job as well as in training (Wigdor and Green, 1991). In addition, project research demonstrated that valid, reliable measures of individual job performance can be developed, including hands-on tests. JPM research supports the use of the ASVAB to select recruits into the military. But if a recruiter wished to predict an individual's performance score for a military job, he or she would

---

[1]The Army did examine second-term job performance in the Career Force Project (e.g., Campbell and Zook, 1992), a follow-up project to Project A.

[2]The Services also developed other performance measures, including less expensive measures such as written tests of job knowledge.

be limited to at best 45 jobs (those having some form of performance criterion) and 33 jobs that have hands-on measures. More desirable would be
the capability to predict an individual's job performance for *any* military
job, whether or not a hands-on measure (or other performance measures)
had been developed for it.

Transporting validation results beyond a specific setting to other settings has been the concern of two methods in the industrial/organizational
psychology literature: validity generalization and synthetic validation. In
this paper, following a brief discussion of these two methods, a third method
that can be used to provide performance predictions for jobs that are devoid
of criterion data—multilevel regression—is introduced and discussed in detail. The application of multilevel regression models to the JPM data is
presented, and results are also given from an investigation of the validity of
the performance predictions derived from the multilevel equations.

VALIDITY GENERALIZATION

For many years, psychologists specializing in personnel selection emphasized the need to demonstrate the validity of selection measures upon
each new application—whether the goal was to predict performance in a
different job, or for the same job in a different setting. The rationale for
this approach was that the validity of a selection measure was specific to
the situation. Indeed, one typically observed a rather large range in the
magnitude of validity coefficients across various situations for the same test
or for similar tests of the same construct (e.g., verbal ability). But conducting job-specific validity studies could be very expensive. Furthermore, for
jobs containing a small number of incumbents, such studies would be likely
to provide either unstable or nonsensical results.

Focusing on this latter shortcoming of the situational specificity hypothesis, Schmidt, Hunter, and their colleagues (e.g., Hunter and Hunter,
1984; Schmidt and Hunter, 1977; Schmidt et al., 1981; Schmidt et al., 1979)
suggested that the majority of variation in observed validity coefficients
across studies could be explained by statistical artifacts. This notion led to
the conclusion that, contrary to the conventional wisdom, the validities of
many selection measures (and cognitive ability measures in particular) were
in fact generalizable across jobs and situations.

In validity generalization, a distribution of validities from numerous
validation studies is created and then corrected for the statistical artifacts of
sampling error, criterion unreliability, and predictor range restriction.[3] The

---

[3]One other correction that has been suggested, correcting for unreliability in the predictors
(Schmidt et al., 1979), should not be used if one wishes to generalize results for observed
predictor measures, taken as they come, rather than for the relationship between true scores on
the predictors and criteria.

result is a new distribution that more accurately reflects the degree of true variation in validity for a given set of predictors. If a large portion of the coefficients in the corrected distribution exceeds a value deemed to be meaningful, then one may conclude the validity generalizes across situations. In essence, validity generalization is a meta-analysis of coefficients from validation studies, and other meta-analytic approaches have been suggested (Hedges, 1988).

Concluding that validity coefficients generalize across situations, however, does not preclude variation in the coefficients across situations. The portion of the corrected distribution lying above the "meaningful" level may exhibit significant variation. If a substantial portion of the variation in the observed coefficients can be attributed to statistical artifacts (75 percent has served as the rule of thumb), however, then the mean of this distribution is considered the best estimate of the validity of the test(s) in question—situational specificity is rejected and the mean value is viewed as a population parameter (i.e., the correlation between the constructs in question). Hunter and Hunter (1984:80) reported that the validity generalization literature clearly indicates that most of the variance in the validity results for cognitive tests observed across studies is due to sampling error such that "for a given test-job combination, there is essentially no variation in validity across settings or time."

Although the findings of validity generalization research have considerably lightened the burden for personnel psychologists interested in demonstrating the validity of certain selection measures, the procedure is not without its critics, and many of its features are questioned (Schmitt et al., 1985; Sackett et al., 1985). Furthermore, the approach does not speak directly to the issue of obtaining performance predictions for jobs devoid of criteria. Although the corrected mean validity could be used to forecast performance scores, the approach is too indirect. Selection decisions are often based on prediction equations, which may in turn comprise a number of tests. The application of validity generalization results to this situation would require a number of validities for the test battery (i.e., composite) in question. Furthermore, even if such results were available, a more desirable approach would be to "focus the across-job analysis on the *regression* parameters of direct interest in the performance prediction, rather than on the *correlations* of the validity generalization analysis" (Bengt Muthén, personal communication, January 18, 1990).

SYNTHETIC VALIDITY

A second alternative to the situational specificity hypothesis is synthetic validity (Lawshe, 1952:32), defined as "the inferring of validity in a specific situation." The basic approach is to derive the validity of some test or test composite by reducing jobs into their components (i.e., behaviors, tasks, or individual attributes necessary for the job) via job analysis, determining the correlations between the components and performance in the job, and aggregating this validity information into a summary index of the expected validity. Wise et al. (1988:78-79) noted that "synthetic validation models assume that overall job performance can be expressed as the weighted or unweighted sum of individual performance components." Mossholder and Arvey (1984:323) pointed out that synthetic validity is not a specific type of validity (as opposed to content or construct validity), but rather "describes the logical process of inferring test-battery validity from predetermined validities of the tests for basic work components."

Several approaches to synthetic validation may be found in the literature, including the J-coefficient (Primoff, 1955) and attribute profiles from the Position Analysis Questionnaire (McCormick et al., 1972). Descriptions of these approaches are available in Crafts et al. (1988), Hollenbeck and Whitemer (1988), Mossholder and Arvey (1984), and Reynolds (1992). To illustrate the process of obtaining synthetic validity estimates, consider the Army's Synthetic Validation Project (SYNVAL) as an example.

The goal of the SYNVAL project (Wise et al., 1991) was to investigate the use of the results of the Army's JPM Project (Project A) to derive performance prediction equations for military occupational specialties (MOS) for which performance measures were not developed. SYNVAL researchers employed a validity estimation task to link 26 individual attributes, which roughly corresponded to the predictor data available on the Project A sample, to job components described in terms of three dimensions (i.e., tasks, activities, or individual attributes). These validity estimates were provided by experienced personnel psychologists.

The validity estimates, in concert with information regarding the importance, difficulty, and frequency of various job tasks or activities (component "criticality" weights) and empirical estimates of predictor construct intercorrelations, were used to generate synthetic equations for predicting job-specific and Army-wide job performance. The task and activity judgments were obtained from subject matter experts (noncommissioned officers and officers). Similar to previous results from a Project A validity estimation study (Wing et al., 1985), the estimates were found to have a high level of interrater agreement. Various strategies for weighting (1) the predictors in the component equations and (2) the component equations to form an overall equation were investigated.

A substantial advantage of the SYNVAL project is the capacity to compare the predicted scores generated by the synthetic equations to existing data. First, synthetic prediction equations were developed for the MOS having performance data. The synthetic equations were compared with ordinary least-squares prediction equations for the corresponding MOS, based on data from Project A validation studies. The data had been corrected for range restriction and criterion unreliability. Validity coefficients for the synthetic equations were found to be slightly lower than the adjusted validity coefficients (adjusted for shrinkage) for the least-squares equations. The differential validity (i.e., the degree to which the validity coefficients for job-specific equations were higher when the equations were applied to their respective job than when applied to other jobs) of the synthetic equations was also somewhat lower than that evidenced by the least-squares equations.

Now that personnel psychologists have provided the estimated linkages between the attributes and Army job components, all that remains is to obtain criticality estimates of each component from subject matter experts for any Army MOS without criteria or for any new MOS. This information can then be combined with the estimated attribute-component relationships to form a synthetic validity equation for the job.

An important characteristic of the synthetic validation approach used in SYNVAL is that synthetic equations could have been developed without any criterion data whatsoever. The presence of actual job performance data, however, allowed validation of the synthetic validity procedure. Although the SYNVAL project demonstrated the synthetic validation technique to approximate optimal (i.e., least-squares) equations closely, some researchers might be wary of deriving equations that rely so heavily on the judgments of psychologists or job incumbents.

A procedure is available that operates directly on the job-analytic information to provide estimated performance equations for jobs without criterion data (although this procedure does require the presence of criterion data for at least a subset of jobs). The procedure is presented by describing its use in the Linkage project (Harris et al., 1991; McCloy et al., 1992), the project that marked the beginning of the enlistment standards portion of the JPM Project. The goal of the Linkage project was to use JPM Project data to investigate the relationship (i.e., linkage) between job performance and enlistment standards. To explore this relationship fully, it was imperative that the method of analysis provide equations that were generalizable to the entire set of military jobs. The synthetic validity approach used in SYNVAL was not selected for the Linkage project because the cost of implementing the approach would have been prohibitive. Specifically, the SYNVAL approach would have necessitated an extensive data collection, because job component information and validity estimates of the predictor/job component linkages were not available for non-Army jobs. In the next section, multilevel regression analysis is proffered as an alternative method for deriving job performance predictions for jobs without criterion data.

MULTILEVEL REGRESSION ANALYSIS

An Example

Suppose that some new selection measures have been developed for predicting performance, and it is of interest to investigate their predictive validity for several jobs. In this example, we have a criterion (e.g., a score from a hands-on test of job performance) P_{ij} for person i in job j. We assume that P_{ij} depends on an individual's aptitude test score (call it A_{ij}; this could be a set of test scores) and some set of other individual character-istics such as education and time in service (call this O_{ij}). We further assume that the effects of these independent variables could differ across jobs and that the jobs are a random sample of the total set of jobs. Thus, the model is

$$P_{ij} = \alpha_j + \beta_j A_{ij} + \gamma_j O_{ij} + \varepsilon_{ij} \qquad (1)$$

where α_j is a job-specific intercept, β_j and γ_j are job-specific slopes, and ε_{ij} is an error term. This is the general form for the equation linking individual job performance to enlistment standards—the linkage equation.

This model says that α_j, β_j, and γ_j can, *in principle*, vary across jobs. Multilevel regression allows one to quantify the variation in these param-eters and to determine if the variation is statistically significant. The varia-tion is addressed by assuming that the parameters themselves have a sto-chastic structure. Namely,

$$\alpha_j = \alpha + a_j, \quad \text{where} \quad a_j \sim N(0, \sigma^2_a) , \qquad (2)$$

$$\beta_j = \beta + b_j, \quad \text{where} \quad b_j \sim N(0, \sigma^2_b) , \qquad (3)$$

$$\gamma_j = \gamma + c_j, \quad \text{where} \quad c_j \sim N(0, \sigma^2_c) , \qquad (4)$$

This formulation says that the intercept for job j (α_j) has two components: α, the mean of all the α_j's (note the lack of the j subscript), and a_j, a component that can be viewed as the amount by which job j's intercept differs from the average job's intercept (i.e., differs from α). Note that the model assumes the distributions of a_j, b_j, and c_j are to be normal; their joint distribution is assumed to be multivariate normal. Although a_j, b_j, and c_j are completely determined for any specific job, the multilevel model con-ceives of these components as random, because the sample of jobs is as-sumed to be chosen at random. If the jobs are picked at random, these components are likewise random. Thus, coefficients modeled to vary across groups (here, jobs) may be labeled *random effects* (indeed, multilevel mod-els are sometimes called random effects models), whereas coefficients modeled to remain constant across groups may be labeled *fixed effects*. The variance components represent the variance of the random effects across jobs. For example, σ^2_a is the variance across jobs of the a_j's and therefore of the α_j's, because α is the same for all jobs.

A multilevel regression model was chosen for the linkage equation because the JPM data are multilevel or "nested." Specifically, in the JPM database, individuals are nested within jobs.[4] Ordinary least-squares (OLS) regression models are inappropriate for multilevel data. To see why this is so, consider a simpler version of equation (2) in which only the intercept (α) is allowed to vary across jobs (i.e., estimate α_j). Thus, the model is

$$P_{ij} = \alpha_j + \beta A_{ij} + \gamma O_{ij} + \varepsilon_{ij} , \tag{5}$$

and α_j is modeled by equation (2). Substituting equation (2) into equation (5) results in a residual term of

$$a_j + \varepsilon_{ij} , \tag{6}$$

implying that the residuals from two individuals in the same job are correlated (i.e., individuals within a job share the same error component, a_j). The same situation obtains for the other parameters as well. Therefore, applying the ordinary regression model to these data would result in biased standard errors for the regression parameters (generally biased downward, increasing the chance of a Type I error; see Green, 1990:478 for more details).

A key feature of the random effects (a_j, b_j, c_j) is that a portion of the variation they represent may be systematic. Consider the general form of the linkage equation given by equation (1). The parameters in this equation possess a j subscript, signifying they may vary across jobs. One might surmise that some of the across-job variation in the parameters may be due to job characteristics (e.g., cognitive demands, demand for psychomotor ability). If so, variables assessing those job characteristics (call them M_j) thought to contribute to this variation could be included in the multilevel regression model. Thus equation (2) would become

$$\alpha_j = \alpha + \pi_\alpha M_j + \eta_{\alpha j}$$

where π_α is a weight applied to the job characteristic variables M_j, and $\eta_{\alpha j}$ is the residual random effect. To the extent that the job characteristic variables were predictive of the parameter variation across jobs, the amount of error in the prediction system would be reduced.

---

[4]The JPM jobs are also nested within Service. The Service level was not modeled, however, because the four Services provide only four observations—an insufficient number of data points to model variation.

Thus, the linkage equation contains not only individual characteristic information, but also variables that assess various characteristics of the military jobs. Before a discussion of the specific form taken by equation (1) in the Linkage project, the variables constituting it deserve comment.

The Building Blocks of the Linkage Equation: Individual- and Job-Characteristic Variables

As noted above, equation (1) gives the general form of the linkage equation. Before discussing the specific form taken by equation (1) in the Linkage project, the variables constituting it deserve comment.

Derivation of the Individual-Characteristic Variables

Measures of individual characteristics were obtained from the Services' JPM data files. The measures were: (1) hands-on job performance test scores (percentage correct),[5] (2) educational attainment (high school diploma graduate or non-high school graduate), (3) experience (total months of service), (4) 10 ASVAB subtest standard scores,[6] and (5) MOS code. The data were available for 8,464 individuals from 24 jobs studied in the JPM Project. A total of 24 MOS were included in the Linkage project:

Army:	N
Infantryman (11B)	663
Cannon crewman (13B)	597
Tank crewman (19E)	465
Single channel radio operator (31C)	346
Light wheel vehicle/power generation mechanic (63B)	594
Motor transport operator (64C)	646
Administrative specialist (71L)	490
Medical specialist (91A)	483
Military police (95B)	657

---

[5]A person's percentage correct score on the hands-on tests does not necessarily indicate the percentage of tasks completed successfully. For example, Army hands-on tests were scored such that each task comprised a number of steps. The score on a task was the percentage of steps the individual performed correctly (scored as "GO"). The average of the task scores was taken as the total hands-on score. Thus, it is incorrect to infer that a score of 90 indicates that the examinee could perform 90 percent of the job tasks correctly. Indeed, one could obtain a score of 90 without ever performing any task entirely correctly. A 90 percent score could be obtained by performing correctly only 90 percent of the steps for each tested task.

[6]The ASVAB subtests are paragraph comprehension (PC), word knowledge (WK), arithmetic reasoning (AR), mathematical knowledge (MK), general science (GS), auto/shop information (AS), electronics information (EI), mechanical comprehension (MC), coding speed (CS), and numerical operations (NO).

Navy:
    Electronics technician (ET)   136
    Machinist's mate (MM)   178
    Radioman (RM)   224
Air Force:
    Aircrew life support specialist (122X0)   166
    Air traffic control operator (272X0)   171
    Precision measuring equipment specialist (324X0)   124
    Avionic communications specialist (328X0)   83
    Aerospace ground equipment specialist (423X5)   216
    Jet engine mechanic (426X2)   188
    Information systems radio operator (492X1)[7]   120
    Personnel specialist (732X0)   176
Marine Corps:
    Rifleman (0311)   940
    Machinegunner (0331)   271
    Mortarman (0341)   253
    Assaultman (0351)   277

The small number of jobs for which performance data were available made reducing the number of predictors in the performance model advisable. Using each of the ASVAB subtest scores as a predictor along with the other measures of individual characteristics and job characteristics would have involved estimating too many parameters, whereas using only the Armed Forces Qualification Test (AFQT) or some other general ability factor might have missed important job differences. The solution to this problem was to use nonoverlapping ASVAB composite scores, thus reducing the number of predictors while retaining as much ability information as possible. The ASVAB composite scores were calculated from a database used to develop the ASVAB-to-Reading Grade Level (RGL) conversion table (Waters et al., 1988) and have been discussed elsewhere (e.g., Campbell, 1986). There are four ASVAB factors:

Quantitative = MK + AR[8]
Speed = NO + CS
Technical = AS + MC + EI
Verbal = PC + WK + GS.

----

[7]When the data for this Air Force specialty were collected, its designation was 293X3— ground radio operator.

[8]The composite scores were calculated using ASVAB subtest standard scores. Each ASVAB subtest standard score had a mean of 50 and a standard deviation of 10.

TABLE 1 Individual Characteristics for the Job Performance
Measurement Project Sample (N = 8,464)

Variable Label	Description	Mean	Standard Deviation
HOPT	Hands-on job performance test score (percentage)	67.37	12.68
AFQT	ASVAB AFQT composite score	206.94	22.22
		(201.03)	(28.07)
TECH	ASVAB technical composite score	159.28	21.16
		(152.57)	(25.04)
TIS	Total months of service	23.65	10.66
EDUC	Educational attainment	1.92	0.28
		(1.73)	(0.48)

The quantitative and verbal factors form the AFQT score:

$$AFQT = PC + WK + MK + AR.[9]$$

To supplement the AFQT, the technical composite was also selected. The speed composite was not used in subsequent modeling efforts because it was not a significant predictor of hands-on performance.

Descriptive statistics were calculated on the individual characteristics for the JPM Project sample (N = 8,464). Means and standard deviations for the hands-on performance (HOPT) test score, the AFQT composite score, the technical composite score (TECH), months of service experience (TIS), and educational attainment[10] (EDUC) are presented in Table 1. Means and standard deviations for AFQT, TECH, and EDUC calculated from the Waters et al. data are in parentheses. Note that the test score distributions for the JPM job incumbents do not differ greatly from the distributions obtained from military recruits. Table 2 contains job-specific means and standard deviations for the individual characteristics for each of the jobs in the JPM Project sample.

Derivation of the Job-Characteristic Variables

Development of the job-level variables for the multilevel model was based on an analysis of civilian jobs. Because job-characteristic data for entry-level military jobs were not available, alternative sources of this type

[9]AFQT excludes the GS subtest.
[10]Educational attainment was coded: 1 = non-high school graduate, 2 = high school graduate.

TABLE 2 Individual Characteristics by Occupation

Occupation	N	Performance Score $(P_{\cdot j})$		AFQT Composite $(A_{\cdot j})$		TECH Composite $(T_{\cdot j})$		Education $(E_{\cdot j})$		Experience $(X_{\cdot j})$	
		Mean	SD	Mean	SD	Mean	SD	Mean	SD	Mean	SD
Army	4,941	70.66	10.48	204.70	22.24	158.74	20.27	1.93	.25	20.70	5.30
Infantryman	663	69.92	7.42	206.01	23.62	162.28	18.66	1.92	.27	20.26	5.51
Cannon Crewman	597	62.03	11.28	196.51	21.70	149.69	21.70	1.90	.30	21.36	5.90
Tank Crewman	465	77.08	7.89	203.99	23.67	162.33	20.02	1.94	.24	19.25	4.78
Radio Operator	346	69.98	7.90	209.06	21.65	158.66	19.90	1.93	.26	20.29	5.46
Vehicle Mechanic	594	84.34	5.03	200.06	21.61	164.00	18.84	1.92	.27	20.08	5.64
Motor Transport	646	71.21	7.67	193.24	21.22	156.70	18.12	1.92	.28	20.32	4.50
Administrative Specialist	490	59.04	8.41	206.43	21.30	142.96	19.29	1.99	.09	22.41	4.56
Medical Specialist	483	71.48	7.25	213.29	17.24	159.35	18.40	1.90	.30	20.89	5.60
Military Police	657	70.22	6.23	216.92	15.50	169.43	15.09	1.97	.18	21.28	4.96
Navy	538	75.71	12.03	211.92	25.97	156.61	23.92	1.97	.17	35.40	14.23
Electronics Technician	136	81.51	9.48	233.98	18.46	176.43	20.34	2.00	.00	43.96	12.86
Machinist's Mate	178	82.16	9.14	217.56	24.40	160.59	20.16	1.95	.22	34.82	14.09
Radioman	224	67.07	9.94	194.04	17.37	141.41	19.88	1.97	.17	30.67	12.74

Air Force	1,244	68.25	13.20	217.73	19.37	165.83	21.75	1.98	.13	28.19	11.46
Aircrew Life Support	166	69.93	13.73	212.63	17.65	157.27	20.34	1.99	.11	28.78	11.10
Air Traffic Control	171	70.08	10.87	227.20	15.03	169.02	19.16	1.98	.13	26.88	8.87
Precision Measuring Equipment	124	76.00	8.84	234.23	13.33	181.83	14.52	2.00	.00	27.57	10.68
Avionic Communications Specialist	83	77.33	11.39	235.24	12.56	184.06	13.72	1.99	.11	35.28	14.85
Aerospace Ground Equipment Specialist	216	57.08	9.51	212.52	17.31	171.85	14.72	1.99	.10	28.04	10.47
Jet Engine Mechanic	188	73.16	10.42	209.47	19.82	174.23	16.41	1.95	.22	29.36	10.78
Information Systems Radio Operator	120	71.06	13.14	212.22	19.21	149.89	21.09	1.98	.13	23.61	13.10
Personnel Specialist	176	61.67	13.66	212.44	17.54	145.47	19.74	1.99	.08	28.04	11.81
Marine Corps	1,741	54.82	9.38	204.01	20.21	156.97	21.41	1.81	.39	25.16	15.25
Rifleman	940	52.62	8.96	201.65	19.61	153.50	21.57	1.79	.41	22.96	12.37
Machinegunner	271	54.79	7.92	203.45	20.52	159.59	20.80	1.85	.36	30.10	18.78
Mortarman	253	52.86	8.76	204.81	21.03	158.30	20.83	1.81	.39	24.10	17.36
Assaultman	277	64.06	6.69	211.84	19.25	165.00	19.37	1.83	.38	28.73	16.47
Total Sample	8,464	67.37	12.68	206.94	22.22	159.28	21.16	1.92	.28	23.65	10.66

of information were sought. A readily available source of job-characteristic information is the *Dictionary of Occupational Titles* (DOT) (U.S. Department of Labor, 1977). The Department of Labor database used to compile the most recent edition of the DOT was obtained from the National Technical Information Service. The database contains the DOT codes for ratings of worker functions and worker traits for approximately 12,000 civilian occupations. Military-civilian job matches were obtained from the Military-Civilian Crosscode Project database (Wright, 1984). Each military job so matched adopted the ratings of its civilian counterpart.

Because the Linkage project was concerned with first-term job performance, only cross-code matches for entry-level jobs were obtained from DoD. The database contained military-civilian equivalents for 965 entry-level military occupations. Once the military-civilian equivalents were obtained, 44 DOT job-characteristic variables describing the civilian jobs were used to characterize the military occupations. These variables represent a variety of job functions and worker traits, including job complexity, training time, aptitude and temperament requirements, physical demands, and environmental conditions.

Job-level composite scores were developed to decrease the number of predictors in the performance equation. To calculate job-level composite scores from the full set of 44 DOT variables, principal components analyses were performed using the 965 military occupations from the cross-code database. An orthogonal component structure was maintained to produce nonoverlapping composites of job characteristics.

A four-factor solution with orthogonal varimax rotation was selected as most appropriate; the four factors accounted for 48 percent of the total variance in the original variables. The first component accounted for 18.5 percent of the variance and consisted of 16 variables that deal mainly with *working with things*, suggesting that this component reflects the extent to which manual labor is a part of the job. The second component accounted for 15 percent of the total variance and consisted of 10 variables that reflect the *cognitive complexity* of work. The third component accounted for 8.6 percent of the variance and included 9 variables dealing with *unpleasant working conditions*. The fourth component accounted for 5.9 percent of the variance and contained 12 items dealing with *fine motor control* and coordination needed in some jobs.

Using the results of the four-component solution, component scores were obtained for the 925 military jobs having complete data (i.e., observations on all 44 occupational variables). These job-specific component scores were then used as the job-characteristic variables in the multilevel regression model of military job performance.[11]

---

[11]See Harris et al. (1991) for a more detailed description of the predictors.

Specification of the Linkage Equation

For the Linkage project, the general specification given in equation (1) is expanded so that job performance is a function of the characteristics of individuals (AFQT standard score, ASVAB technical composite score, time in service, and education). Job performance, in turn, is operationalized as scores on the job-specific hands-on performance tests developed for each service during the JPM Project. Thus, the prediction equation takes the form:

$$P_{ij} = \alpha_j + \beta_j A_{ij} + \phi_j T_{ij} + \gamma E_{ij} + \delta_j X_{ij} + \rho T_{ij} X_{ij} + \varepsilon_{ij} \tag{7}$$

where P_{ij} is the hands-on performance test (HOPT) score for person i in job j; A_{ij} is the AFQT composite score; T_{ij} is the ASVAB technical composite (TECH) score; E_{ij} is educational attainment (EDUC); X_{ij} is time in service (TIS); and α_j, β_j, ϕ_j, γ, δ_j, and ρ are model parameters. Note that γ has no j subscript because the effect of education was found not to vary across jobs:

$$\gamma_j = \gamma. \tag{8}$$

Structure of the Linkage Parameters

The structure of the model parameters for the linkage equation is the following:

$$\alpha_j = \alpha + \pi_\alpha M_j + \eta_{\alpha j}, \tag{9}$$

$$\beta_j = \beta + \pi_\beta M_j + \eta_{\beta j}, \tag{10}$$

$$\phi_j = \phi + \pi_\phi M_j + \eta_{\phi j}, \tag{11}$$

$$\delta_j = \delta + \pi_\delta M_j + \eta_{\delta j}, \tag{12}$$

where α, β, δ, and ϕ are the mean values of the parameters across all jobs (note the lack of the j subscript), the π's are vectors of coefficients constrained to be the same across jobs (i.e., they are "fixed" coefficients), M_j is a vector of four standardized component scores that describe job characteristics (working with things, cognitive complexity, unpleasant working conditions, and fine motor control) described earlier, and the η's are random variation.[12] (To generalize the model to the universe of first-term military

---

[12]In this multilevel parameter specification, the job-level variables do not need to be the same for all parameters, although they are in the linkage equation. In addition, the random error terms may covary; this covariation is estimated in our model.

jobs, the job-level coefficients—the π's—cannot be job-specific.) Because the linkage equation contains both individual- and job-level predictors, it qualifies as a multilevel model, with individuals being level one and jobs being level two.[13]

This structure for the model parameters assumes that some of their variation (except for γ) is due to characteristics of the jobs. The M_j variables represent characteristics of jobs believed to influence an individual's performance. The inclusion of such job-characteristic information is our attempt to generalize from our small sample of jobs (the 24 JPM jobs having hands-on criterion data) to the population of military jobs. To the extent that some portion of the parameter variation is due to job characteristics and the proper job-characteristic variables (M_j's) are included in the multilevel model, the amount of variance in the parameters that is unaccounted for can be reduced, thereby increasing the accuracy of prediction or, equivalently, decreasing the standard error of estimate.

The M_j variables reduce the uncertainty in the job-specific parameters by absorbing some of the variation across jobs that would be part of the random effect if the M_j variables were not in the model. For example, for the job-specific intercept α_j, the term $\pi_\alpha M_j$ models part of the variation in intercept parameters across jobs that otherwise would be part of the random effect a_j. Including the second-level variables should reduce the uncertainty in the estimation of the α_j's. This same logic holds for all other model parameters.

The multilevel model may be approximated by a fixed-effects (i.e., conventional OLS) regression model. Substituting equations (8)-(11) into (7) gives the following:

$$P_{ij} = (\alpha + \pi_\alpha M_j + \eta_{\alpha j}) + (\beta + \pi_\beta M_j + \eta_{\beta j})\,A_{ij} + (\phi + \pi_\phi M_j + \eta_{\phi j})T_{ij} \\ + \gamma E_{ij} + (\delta + \pi_\delta M_j + \eta_{\delta j})X_{ij} + \rho T_{ij}\,X_{ij} + \varepsilon_{ij}\,. \tag{13}$$

Multiplying through and collecting terms yields:

$$P_{ij} = \alpha + \beta A_{ij} + \phi T_{ij} + \gamma E_{ij} + \delta X_{ij} + \rho T_{ij} X_{ij} \\ + (\pi_\alpha M_j + \pi_\beta M_j A_{ij} + \pi_\phi M_j T_{ij} + \pi_\delta M_j X_{ij}) + Z\,, \tag{14}$$

where

$$Z = \eta_{\alpha j} + \eta_{\beta j} A_{ij} + \eta_{\phi j} T_{ij} + \eta_{\delta j} X_{ij} + \varepsilon_{ij}\,. \tag{15}$$

Thus, a model containing the job-characteristic variables and interactions between the job-characteristic variables and the variables whose effects are

---

[13]Those familiar with analysis of variance will recognize this as a mixed model—one having both random and fixed effects.

to vary across occupations may be used to estimate the structural parameters (regression coefficients) in the multilevel analysis. The standard errors of the parameter estimates for this model will be biased, however, due to the failure of the fixed-effects regression to adequately model the correlations among errors in the multilevel error structure. The standard errors will typically be smaller than they should be, thereby increasing the probability of a Type I error.

To determine if the simpler fixed-effects approximation adequately characterized the Linkage data, equation (14) was estimated and compared with the multilevel model. The two sets of parameter estimates were sufficiently different to suggest retaining the multilevel model (see Harris et al., 1991, for a detailed description of the formulation of the performance equation).

The Parameters for the Linkage Equation

The linkage equation's parameters were estimated using the VARCL software package for the analysis of multilevel data (Longford, 1988). This program uses maximum likelihood estimation to obtain parameter values for the model. The specific, unstandardized parameter estimates for the linkage equation are given in Table 3, along with their associated standard errors. (Table 1 contains the means and standard deviations of the individual-characteristic variables.) Also included in this table are values giving the square root of the variance (designated σ) of each parameter across jobs (i.e., the random effects remaining after taking into account job-characteristic information) and their respective standard errors (SE_σ).

The standard errors of the fixed-effect parameter estimates (e.g., β, ϕ) indicate that all of these values are significant. Thus, the variables in the linkage equation demonstrate statistically reliable predictive relationships with job performance for the total sample of jobs. The SE_σ values indicate that the σ values for the random-effect parameters are also significant, suggesting reliable variation remains in these parameters across the sample of jobs, even after the job-characteristic variables are used to model this variation.

Examination of the standard errors for the job-level parameter estimates (e.g., $\pi_{\alpha 1}$, $\pi_{\delta 3}$) shows how well the four job-characteristic variables model the across-job variation in the parameter values for each variable. Specifically, the factor scores do account for a statistically significant portion of the variation in the TECH parameter but do not do as well for the remaining parameters, the values of the intercept term being of particular note (i.e., $\pi_{\alpha 1}$ through $\pi_{\alpha 4}$). Combined with the results of the variation in the random effects (the σ values), we conclude that there might be other moderators (i.e., M_j variables) that would better model the variation of the parameters of the primary linkage equation.

TABLE 3 Coefficient and Variance Component Estimates for the Primary Linkage Equation: the INTERCEPT, AFQT, TECH, and TIS as Random Effects with the EDUC and TECH × TIS Interaction as Fixed Effects

Coefficient		Estimate	Error Standard	σ	SE_{σ}
Intercept	α	31.434		6.679	.974
FS$_1$	$(\pi_{\alpha 1})$	2.763	4.607		
FS$_2$	$(\pi_{\alpha 2})$	5.700	4.086		
FS$_3$	$(\pi_{\alpha 3})$	3.779	4.001		
FS$_4$	$(\pi_{\alpha 4})$	−1.591	8.185		
AFQT	β	.074	.014	.039	.009
AFQTFS1	$(\pi_{\beta 1})$	−.030	.014		
AFQTFS2	$(\pi_{\beta 2})$	.001	.012		
AFQTFS3	$(\pi_{\beta 3})$	−.020	.012		
AFQTFS4	$(\pi_{\beta 4})$	−.036	.025		
TECH	ϕ	.110	.017	.035	.008
TECHFS1	$(\pi_{\phi 1})$	.035	.013		
TECHFS2	$(\pi_{\phi 2})$	−.006	.011		
TECHFS3	$(\pi_{\phi 3})$	.023	.011		
TECHFS4	$(\pi_{\phi 4})$	.050	.023		
EDUC	γ	.882	.325		
TIS	δ	.421	.074	.079	.016
TISFS1	$(\pi_{\delta 1})$	.032	.029		
TISFS2	$(\pi_{\delta 2})$	−.048	.025		
TISFS3	$(\pi_{\delta 3})$	−.082	.024		
TISFS4	$(\pi_{\delta 4})$	−.016	.049		
TECHxTIS	ρ	−.001	.000		

Take note that the application of multilevel analysis in the Linkage project is rather atypical. That is, most applications of multilevel regression analysis involve many groups (e.g., 100-200 schools) with relatively few members in each group (e.g., 20-30 students). In the present analyses, there are relatively few groups (i.e., 24 jobs) that contain many members (sample sizes ranging from 83 to 940). Although this complicates estimation of the variance components for the job-level parameters (i.e., the σ's in Table 3), these components are estimated with enough precision to be statistically significant.

The VARCL output also includes a covariance matrix of the random effects. This matrix provides information regarding the degree to which the parameter values across jobs for one variable covary with the parameter values across jobs for another variable. The covariances for the linkage parameters are presented in Table 4, along with the corresponding correlations. As an example of the information provided in the table, the substantial negative correlation between the intercept and TECH indicates that the TECH parameter tends to be smaller in jobs having a higher overall mean performance level.

TABLE 4 Covariances and Correlations Among the Random Effects

	Intercept	AFQT	TECH	TIS
Intercept	**44.604**	−1.313	−1.267	−1.437
AFQT	−.50	**.002**	−.003	.006
TECH	−.55	−.19	**.001**	−.007
TIS	−.27	.20	−.26	**.006**

Note: Variances appear on the diagonal, covariances ($\times 10^{-1}$) above the diagonal, correlations below the diagonal.

Variance of Predicted Performance Scores

Although VARCL provides standard errors of the model parameters, no standard error of estimate is printed. One may be calculated, however, by taking the square root of the equation:

$$V\left(\hat{P}_{ij}\right) = V\left(\hat{\alpha}\right) + A^2 V\left(\hat{\beta}\right) + T^2 V\left(\hat{\phi}\right) + E^2 V\left(\hat{\gamma}\right) + X^2 V\left(\hat{\delta}\right) + T^2 X^2 V\left(\hat{\rho}\right)$$
$$+ 2A \operatorname{cov}\left(\hat{\alpha}, \hat{\beta}\right) + 2T \operatorname{cov}\left(\hat{\alpha}, \hat{\phi}\right) + 2X \operatorname{cov}\left(\hat{\alpha}, \hat{\delta}\right)$$
$$+ 2AT \operatorname{cov}\left(\hat{\beta}, \hat{\phi}\right) + 2AX \operatorname{cov}\left(\hat{\beta}, \hat{\delta}\right) + 2TX \operatorname{cov}\left(\hat{\phi}, \hat{\delta}\right)$$

(16)

where $V(\bullet)$ and $Cov(\bullet, \bullet)$ are the variance and covariance of the random effects, respectively. The terms A, T, E, and X are the mean AFQT, TECH, EDUC, and TIS values for the job under consideration (see Harris et al., 1991). Note that no terms including job-characteristic variables appear in equation (16). This is because the parameters of the job-characteristic variables (the π's) are fixed (the values of the job-characteristic variables are constants for a given job).

The information for this equation is available from Tables 3 and 4. There are two sources of variability in the job-specific parameters. The largest is the variance of the coefficients across jobs (σ^2_{α}, σ^2_{β}, σ^2_{ϕ}, σ^2_{δ}), the square roots of which are given in Table 3. This source arises because the coefficient α_j is imperfectly estimated by ($\alpha + \pi_{\alpha} M_j$), β_j by ($\beta + \pi_{\beta} M_j$), and so on. Residual error variance (i.e., the η's), conditional on the job-characteristic variables, remains in the parameter estimates. The variance of these errors across the population of jobs is the variance component, $\sigma^2_{\bullet}$.

The second (and usually smaller) source of variation is the variance of the estimate of the mean effects (e.g., $\hat{\beta}$, $\hat{\phi}$). These are the standard errors of the parameters given in Table 3. Thus, for the TECH parameter $\hat{\phi}_j$,

$$V\left(\hat{\phi}_j\right) = \sigma_\phi^2 + \left[SE\left(\hat{\phi}\right)\right]^2 \tag{17}$$

where σ_ϕ^2 is the variance component for the slope ϕ_j and $SE(\hat{\phi})$ is the standard error of $\hat{\phi}$ (the estimate of the mean of the ϕ_j's). The intercept contributes only its random effects component to the variability of $\hat{\alpha}_j$, because a shift in the mean of the intercepts does not contribute to variability in the $\hat{P}_{ij}$ values. Slopes that are constrained to be the same across all jobs (i.e., are a fixed, but not a random, effect) contribute only their $[SE(\cdot)]^2$ component. Thus, for the EDUC parameter,

$$V\left(\hat{\gamma}\right) = \left[SE\left(\hat{\gamma}\right)\right]^2 . \tag{18}$$

The covariance terms in equation (16) are the covariances between the conditional (or residual) random effects (the random effects not accounted for by the second-level variables—the job characteristics). These values are given in Table 4. Note that no covariances are given between the random parameters that vary across jobs and the parameters constrained to be the same across jobs (i.e., the fixed parameters γ and ρ). Because the fixed parameters do not vary across jobs, their covariance terms equal zero.

Table 5 contains the means of the predicted performance scores and associated standard errors of estimate for the 24 JPM jobs. The values of the standard error range from 6.80 for Army MOS 64C (motor transport operator) to 9.62 for the Navy Rating ET (electronics technician). In no instance does a 95-percent confidence interval around the predicted scores yield an implausible value (i.e., less than zero or greater than 100).

Much of the variation in prediction is due to scaling differences across jobs (specifically, intercept variation introduced by the differences in mean HOPT scores across jobs). The HOPT scores were not standardized within jobs but rather remain in their original metric. By retaining the original HOPT metric, the mean differences in HOPT scores across jobs remain. Part of the mean differences in performance scores is attributable to differences in job difficulty. Another contributor to the mean differences in HOPT across jobs is differences in test difficulty, with the difficulty of a given hands-on test being determined by comparing that test to other tests that could be constructed *for a given job*. For example, recall that the Marine Corps JPM sample has a wider range of experience than the Army. The Marine Corps HOPT for rifleman (the Marine Corps's infantry MOS) contains items that assess performance of second-tour tasks. Clearly, this test would be more difficult for first-term soldiers than a test assessing

TABLE 5 Standard Errors of Predicted Performance
Scores

MOS	Predicted Performance Score	Standard Error
11B	62.01	6.32
13B	59.87	6.03
19E	64.29	6.23
31C	70.45	6.45
63B	74.30	6.08
64C	59.26	5.86
71L	63.45	6.48
91A	64.70	6.62
95B	69.67	6.73
ET	79.29	8.11
MM	75.14	7.12
RM	69.67	6.19
122X0	66.73	6.76
272X0	69.87	7.25
324X0	76.41	7.52
328X0	79.70	7.80
423X5	69.82	6.69
426X2	79.58	6.61
492X1	71.09	6.68
723X0	64.64	6.80
0311	60.84	6.23
0331	63.97	6.43
0341	63.15	6.35
0351	64.60	6.70

performance on tasks that are performed during the first tour. In contrast, the Army's infantry HOPT assesses performance on first-term tasks only. This difference in performance test design is reflected in the mean HOPT scores for Marine Corps rifleman (mean = 52.62, SD = 8.96) and Army infantryman (mean = 69.92, SD = 7.42) given in Table 2.

Job difficulty variance is desirable to retain, but test difficulty variance is not. Although standardizing within job would remove the unwanted test difficulty differences, the desired job differences would also be lost. For the approach of leaving the scores in their original metric to be tenable, one must assume that the variance due to test difficulty across jobs is uncorrelated with the individual- and job-characteristic variables. Considering the job-characteristic variables, there is no reason to believe that the characteristics of the job and the difficulty of the test are related. It is possible that more difficult jobs have more difficult tests. If this is due to the content of the job, however, then this reflects job difficulty, not test difficulty, and is not a concern. Thus, for any given job, tests may be relatively easy or relatively difficult. For example, there is no reason to suspect that tests for jobs requiring performance in difficult working conditions would be easier or more difficult than tests for other jobs. Test comparisons as difficult or easy must be made within a job. Comparisons across jobs are confounded by job difficulty.

The reasoning for a negligible correlation between test difficulty and individual-characteristic variables runs parallel to that for job characteristics. Although intelligent people are placed in difficult jobs, this relationship is based on job content, not test difficulty. There is no reason to believe that people with certain individual characteristics are disproportionately placed into jobs that have very difficult or very easy tests—again, difficult or easy represents a relative comparison within a job rather than across jobs.

Deriving Job-Specific Equations

As mentioned above, the linkage to be made in the Linkage project was the relationship between measures of recruit quality (i.e., enlistment standards) and job performance. The multilevel performance equation given in equation (7) characterizes this relationship for the 24 JPM occupations. In addition, this prediction model is the primary linkage equation—it is the progenitor of all job-specific linkage equations. That is, this model's parameter estimates (given in Table 3) are used to calculate the parameters for equations that allow job-specific performance predictions.

The principal advantage of the primary linkage equation is that it allows performance predictions for jobs having no criterion data. Using ordinary regression, performance scores can be estimated for individuals without criterion data by weighting their predictor information by the appropriate regression coefficients. Performance data, however, are needed for some individuals in that job before the job-specific equation may be estimated. By including job-characteristic variables in our multilevel model, job-specific parameters can be derived for any job having job-characteristic data. These parameters are functions of the job-characteristic variables and, together with the fixed effects of EDUC and the interaction between TECH and TIS, constitute job-specific linkage equations.

Returning to Table 3, the value associated with a Greek letter represents the mean effect of the variable across jobs (e.g., $\beta = .074$). The parameters subscripted from 1 to 4 (e.g., $\pi_{\beta j}$) signify the values in the π vector that are applied to the four component scores, respectively. For AFQT, these values are $-.030$, $.001$, $-.020$, and $-.036$. Substituting these values into equations (8) through (11) allows the estimation of job-specific parameters. Equations (8) through (11) also demonstrate that these estimated job-specific parameter estimates are deviations from the mean parameter estimate—the degree of deviation being a function of the job's factor scores. (Note that γ and ρ, fixed across jobs, do not have any corresponding subscripted values.) Consider the Army MOS 11B (infantryman). The four factor scores for this MOS are -0.68, -2.41, 2.33, and 0.18. Substituting these M_j values and the multilevel parameter estimates just given into equation (9) yields the AFQT parameter (β_j) for predicting job performance for infantrymen:

$$
\begin{aligned}
\beta j &= \beta + \pi_\beta M_j + \eta_{\beta j} \\
&= .07 + [(-.030)(-.68) + (.001)(-2.41) + \\
&\quad\quad (-.020)(2.33) + (-.036)(-.32)] \\
&= .07 + (-.03) \\
&= .04.
\end{aligned}
\tag{19}
$$

Substituting into equations (8), (10), and (11), the same procedure yields the remaining parameters for MOS 11B:

$$
\begin{aligned}
\alpha_j &= 24.35 \\
\phi_j &= .16 \\
\delta_j &= .32.
\end{aligned}
\tag{20}
$$

The coefficients for the job-specific linkage equations for the 24 JPM jobs used in the Linkage project are given in Table 6.

The same procedure affords job-specific parameters for jobs without criterion data. For example, continuing with the AFQT parameter, the following job-specific value is obtained for the Army MOS combat engineer (12B) using its factor scores of $-.51$, -3.09, 1.90, and $-.97$:

$$
\begin{aligned}
\beta j &= \beta + \pi_\beta M_j + \eta_{\beta j} \\
&= .07 + [(-.030)(-.51) + (.001)(-3.09) + \\
&\quad\quad (-.020)(1.90) + (-.036)(-.97)] \\
&= .07 + .02 \\
&= .09,
\end{aligned}
\tag{21}
$$

and for the other parameters,

$$
\begin{aligned}
\alpha_j &= 21.13 \\
\phi_j &= .13 \\
\delta_j &= .43.
\end{aligned}
\tag{22}
$$

Note that the value for β and the four π_β values remain constant in the β_j equations for both MOS; the equations differ only in the M_j values.

Performance equations may also be generated for *groups* of jobs. For example, jobs were grouped into 9 of the 10 DoD occupation codes (see

TABLE 6   Regression Coefficients for Job-Specific Linkage Equations for the 24 JPM Jobs Used in the Linkage Project

MOS	Intercept (α_j)	AFQT (β_j)	TECH (ϕ_j)	TIS (δ_j)	EDUC (γ)	TECH×TIS (ρ)
11B	24.35	0.04	0.16	0.32	.88	−.001
13B	22.17	0.05	0.15	0.39	.88	−.001
19E	25.42	0.04	0.16	0.39	.88	−.001
31C	30.74	0.09	0.09	0.47	.88	−.001
63B	37.90	0.04	0.15	0.34	.88	−.001
64C	19.09	0.07	0.13	0.45	.88	−.001
71L	22.35	0.14	0.03	0.48	.88	−.001
91A	29.40	0.05	0.14	0.26	.88	−.001
95B	31.64	0.11	0.07	0.27	.88	−.001
ET	29.08	0.04	0.16	0.50	.88	−.001
MM	34.72	0.05	0.13	0.43	.88	−.001
RM	32.75	0.07	0.11	0.40	.88	−.001
112X0	22.48	0.08	0.11	0.51	.88	−.001
272X0	27.60	0.11	0.07	0.38	.88	−.001
324X0	33.37	0.05	0.14	0.41	.88	−.001
328X0	34.64	0.05	0.13	0.48	.88	−.001
423X5	26.64	0.04	0.16	0.44	.88	−.001
426X2	41.41	0.02	0.17	0.34	.88	−.001
492X1	30.74	0.09	0.09	0.47	.88	−.001
732X0	18.09	0.13	0.06	0.51	.88	−.001
0311	24.35	0.04	0.16	0.32	.88	−.001
0331	23.73	0.04	0.17	0.34	.88	−.001
0341	23.73	0.04	0.17	0.34	.88	−.001
0351	27.19	0.05	0.14	0.31	.88	−.001

Table 7) in the Linkage project.[14]   Scores on the four factors for the nine job groups were obtained by calculating the weighted (by sample size) mean of the factor scores across the jobs in each of the nine occupational codes. As above, the weighted mean factor scores would be inserted into the primary linkage equation to generate performance equations for each of the occupational codes.

The model also may be amended to include additional or different individual and job characteristics. All that is required is to reestimate the primary linkage equation with the new variables in the model so that new parameter values may be obtained. The procedure just described still applies.

---

[14]No jobs were grouped into occupation code 9, which contains students, patients, trainees, and others not occupationally qualified.

TABLE 7 Job Categories Represented by the DoD
Occupation Codes

Code	Job Category
0	Infantry, gun crews, and seamanship specialists
1	Electronic equipment repairmen
2	Communications and intelligence specialists
3	Health care specialists
4	Other technical and allied specialists
5	Functional support and administration
6	Electrical/mechanical equipment repairmen
7	Craftsmen
8	Service and supply handlers
9	Nonoccupational

Validation of the Primary Linkage Equation

The capacity to generate prediction equations for jobs without criterion data (given that job-characteristic data are available) is a very attractive feature of the multilevel primary linkage equation. Nevertheless, at least two issues should be addressed with regard to the ability of the primary equation to generate job-specific linkage equations that yield quality predictions for jobs without criterion data.

First, the primary equation was estimated on a sample of only 24 jobs. Although these jobs have various desirable qualities—for example, they are high-density jobs, and they are representative in the sense that they span important job groups in the Services (e.g., mechanical, administrative, and combat occupations)—there are still several limitations inherent in them for the Linkage project. For example, they do not span the job-characteristic space defined by the four component scores (e.g., most of the jobs are low in cognitive complexity), and 24 jobs is not a large number of cases for estimating across-job variability. As a result, there is some question about the degree to which the parameters from the primary linkage equation and the corresponding job-specific equations would change if any one of the 24 jobs were deleted from the sample.

Second, quite apart from the capability simply to generate job-specific linkage equations for jobs devoid of criterion data and the independence of those equations from the 24 jobs included in the estimation sample is the issue of how well those generated equations actually predict performance in the out-of-sample jobs. Such information is crucial for evaluating the validity of the performance equation. Hence, there are two primary issues to be addressed: (1) the sensitivity of the primary linkage equation to the jobs in the estimation sample and (2) the validity of the job-specific linkage equations that are generated by the primary linkage equation.

Parameter sensitivity results were reported by McCloy et al. (1992). They estimated the linkage equation 24 times, once as each job in the sample was removed from consideration. Distributions of the parameters from the resulting 23-job equations were obtained, and ratios of predicted performance scores were calculated both across AFQT categories within jobs and within AFQT category across jobs. They found that (1) the parameters were not unduly influenced by the presence of any particular JPM job in the sample of 24 and (2) the predicted performance scores evidenced reasonable stability, suggesting that the performance equations yield consistent results and are mostly unaffected by the presence or absence of specific jobs in the estimation sample.

The second question refers to the validity of the job-specific linkage equations generated by the primary linkage equation. The results of the validity analyses are given here. The capacity to generate predicted performance scores (via job-specific linkage equations) for individuals in jobs for which no criterion data are available begs the question of how well the job-specific linkage equations predict performance for various jobs—in particular, any out-of-sample jobs without criterion data.

Answering this question requires jobs that have criterion data but were not part of the estimation sample for the primary linkage equation. There are essentially two ways such a situation could arise: (1) manufacture such a situation out of the extant sample by using a holdout procedure or (2) obtain relevant data on one or more new jobs after estimating the original primary linkage equation. Both conditions obtained in the present analyses.

Method

The primary linkage equation can be used to generate a job-specific equation for any job having job-characteristic data, whether it appears in the estimation sample or not. The quality of the predictions from any of these equations is of interest, but perhaps the most stringent test of the linkage methodology lies in the prediction of performance scores for out-of-sample jobs. To investigate the validity of the job-specific linkage equations, two types of analyses were performed.

First, each of the 24 JPM jobs was held out of the sample and a "reduced" primary equation was estimated on the remaining 23 jobs. This process resulted in 24 reduced equations. These reduced equations were used to generate a job-specific equation for their corresponding holdout (i.e., out-of-sample) job. The existence of criterion data for each holdout job permitted the observed performance scores for each job to be correlated with the performance scores predicted by the corresponding job-specific linkage equation.

Second, job-specific linkage equations were generated from the 24-job primary linkage equation for two Navy ratings [electrician's mate (EM) and gasoline turbine mechanic (GSM)] and five Marine Corps jobs [organizational automotive mechanic (3521), helicopter mechanic CH-46 (6112), helicopter mechanic CH-53 (6113), helicopter mechanic U/AH-1 (6114), and helicopter mechanic CH-53E (6115)] that were not part of the original estimation sample. These equations yielded predicted performance scores for the individuals in the additional Navy jobs. As in the holdout analyses, the correlation between the observed and predicted performance scores was obtained.

When conducting cross-validation, one typically splits the total sample into two random subsamples, developing a prediction equation on the first sample and applying that equation to the second sample. But this method underestimates the R^2 expected if the regression equation were developed using the entire sample and then applied to the population, because the best set of regression weights (i.e., the full-sample weights) is not used (Campbell, 1990). The present analyses do not match typical cross-validation procedures exactly: job-specific equations were (1) generated from a primary linkage equation that in turn was estimated from a sample of 23 (holdout analyses) or 24 (new Navy and Marine Corps specialties) jobs and (2) applied to their respective jobs that were not part of the estimation sample. That is, the validation sample was a *job* that was not part of the original sample of 23 (holdout analyses) or 24 (new Navy and Marine Corps specialties) jobs, rather than a random subsample of the original sample. Nevertheless, both sets of analyses provide an empirical test of the ability of the job-specific linkage equations to provide accurate predictions of actual performance scores for out-of-sample jobs. Such information is vital because these situations reproduce the scenario in which the primary linkage equation would be implemented by manpower planners.

Results

The results of the analyses are presented in Table 8, which contains (1) the sample size for each job (N), (2) the squared multiple correlation for the least-squares job-specific regression equations (R^2_{OLS}), (3) the squared multiple correlation for the job-specific linkage equation generated from the reduced 23-job "primary" equation (R^2_{cv}) (i.e., the correlation between the predicted performance scores taken from the job-specific linkage equation and the actual observed performance scores), (4) the difference between the values of R^2_{OLS} and R^2_{cv}, and (5) R^2_{OLS} values adjusted using various shrinkage formulae (R^2_{adj}).

Two features of the first two columns of R^2 values are of note: (1) the values are quite variable, ranging from .065 to .508 for R^2_{OLS} and .031 to .461 for R^2_{cv}, and (2) $R^2_{OLS} > R^2_{cv}$. The latter finding is expected, given that

TABLE 8 R^2 Obtained for the Job-Specific Least-Squares and Linkage Equations[a] and Shrinkage Expected Using 4 Formulae

JOB	N	R$^2_{OLS}$	R$^2_{cv}$	Difference[b]	Shrinkage Formula (R$^2_{adj}$)			
					Wherry 1931	Browne 1975[c]	Rozeboom 1978	Lord 1950-Nicholson 1960
11B	663	0.086	0.080	0.006	0.080	0.076	0.075	0.073
13B	597	0.065	0.038	0.027	0.059	0.054	0.052	0.051
19E	465	0.141	0.126	0.015	0.133	0.128	0.126	0.124
31C	346	0.140	0.103	0.037	0.130	0.123	0.120	0.117
63B	594	0.076	0.057	0.019	0.069	0.065	0.063	0.062
64C	646	0.108	0.089	0.019	0.103	0.099	0.097	0.096
71L	490	0.127	0.110	0.017	0.120	0.114	0.112	0.111
91A	483	0.117	0.056	0.061	0.110	0.104	0.102	0.100
95B	657	0.102	0.056	0.046	0.097	0.093	0.091	0.090
ET	136	0.056	0.025	0.053	0.039	0.025	0.018	0.081
MM	178	0.154	0.120	0.035	0.135	0.122	0.116	0.111
RM	224	0.154	0.099	0.054	0.138	0.128	0.123	0.119
EM	80	0.348	0.281	0.067	0.313	0.288	0.279	0.270

GSM							
GSM	88	0.140	0.077	0.063	0.098	0.076	0.058	0.047
112	166	0.141	0.106	0.034	0.119	0.105	0.098	0.093
272	171	0.077	0.031	0.046	0.055	0.043	0.033	0.027
324	124	0.224	0.181	0.042	0.198	0.180	0.172	0.165
328	83	0.223	0.140	0.083	0.183	0.157	0.144	0.134
423	216	0.173	0.149	0.024	0.157	0.146	0.142	0.138
426	188	0.088	0.050	0.038	0.068	0.056	0.049	0.043
492	120	0.216	0.178	0.038	0.189	0.170	0.162	0.155
732	176	0.226	0.198	0.028	0.208	0.195	0.190	0.185
031	940	0.324	0.314	0.010	0.321	0.319	0.318	0.317
033	271	0.358	0.297	0.061	0.348	0.341	0.338	0.336
034	253	0.379	0.366	0.013	0.369	0.362	0.360	0.357
035	277	0.238	0.230	0.008	0.227	0.219	0.216	0.213
3521	907	0.240	0.176	0.064	0.237	0.233	0.234	0.232
6112	152	0.464	0.461	0.003	0.449	0.435	0.438	0.431
6113	93	0.508	0.453	0.055	0.486	0.464	0.469	0.458
6114	190	0.187	0.167	0.020	0.169	0.152	0.157	0.148
6115	113	0.237	0.203	0.034	0.209	0.181	0.189	0.174

[a] All job-specific equations derived from 23-job performance equations except EM, GSM, 3521, 6112, 6113, 6114, and 6115 (derived from the 24-job equation).

[b] Difference = $R^2_{OLS} - R^2_{CV}$.

the least-squares equations are optimal for the samples on which they were derived; the job-specific linkage equations are not. The largest differences between R^2_{OLS} and R^2_{cv} primarily occur in the jobs having the smallest sample sizes (e.g., EM, GSM, 328X0). The absolute magnitude of the differences is not particularly large, however, ranging from .006 for 11B to .083 for 328X0. The question remaining is what to make of this difference in R^2 values.

Shrinkage Formulae

The value of R^2 obtained for the least-squares job-specific regression equation (R^2_{OLS} in Table 8) can be viewed as an upper bound because calculating least-squares regression weights capitalizes on chance fluctuations specific to the sample in which the equation is developed. Applying the weights from this equation to another sample would result in a decrease in R^2, because the weights are suboptimal for the second sample. Thus, the R^2 yielded by the regression weights "shrinks" relative to the original R^2.

The amount of shrinkage to be expected may be estimated using a shrinkage formula. Perhaps the best known of these is a formula developed by Wherry (1931):

$$R^2_{adj} = 1 - \left(\frac{N-1}{N-k-1}\right)\left(1 - R^2_{yx}\right) \tag{23}$$

where N is the size of the sample used to estimate the equation, k is the number of predictors, and R^2_{yx} is the sample coefficient of determination (R^2_{OLS} from Table 8). Wherry's formula gives the value for R^2 expected if the equation were estimated in the population rather than a sample.

Because the population will virtually never be at the researcher's disposal, Wherry's formula is of little practical value. As noted by Darlington (1968) and Rozeboom (1978), the Wherry formula does not answer the more relevant question of what the R^2 would be if the *sample* equation were applied to the population. Both Cattin (1980) and Campbell (1990) reported that no totally unbiased estimate for this value exists, although the amount of bias inherent to current estimates is generally small. They recommended a formula developed by Browne (1975), on the basis of its desirable statistical properties. Browne's formula, appropriate when the predictor variables are random (as opposed to fixed), is

$$R^2_{adj} = \frac{(N-k-3)\rho^2 + \rho}{(N-2k-2)\rho + k} \tag{24}$$

where ρ is the adjusted R^2 from the Wherry formula; N and k are defined as above. In truth, Browne's formula contains two terms, this equation being the first (and by far the larger). Browne reported the bias introduced by neglecting the second term of his R^2 adjustment to be no greater than .02. (He also provided an equation for fixed predictor variables.)

A second formula for estimating the validity of the sample equation in the population was provided by Rozeboom (1978):

$$R^2_{adj} = 1 - \left(\frac{N+k}{N-k}\right)\left(1 - R^2_{yx}\right) \tag{25}$$

with N, k, and R^2_{yx} defined as above.

The shrinkage formulae just described allow one to estimate the population multiple correlation for the full sample equation. If the average sample cross-validity coefficient is of interest, Lord (1950) and Nicholson (1960) independently developed a shrinkage formula for estimating this value:

$$R^2_{adj} = 1 - \left(\frac{N+k+1}{N}\right)\left(\frac{N-1}{N-k-1}\right)\left(1 - R^2_{yx}\right) \tag{26}$$

with N, k, and R^2_{yx} defined as above.

Comparison of Adjusted and Cross-Validity R^2 Values

Because the job-specific least-squares equations are optimal for the samples on which they were developed but the job-specific linkage equations are not, the comparison of R^2_{OLS} to R^2_{cv} is not exactly fair. A more equitable comparison obtains through adjustment of the R^2_{OLS} values for shrinkage. Thus, the four shrinkage formulae were applied to the R^2 values from the least-squares job-specific regression equations (i.e., R^2_{OLS}). These adjusted R^2 values (R^2_{adj}) were then compared to the cross-validity R^2 values obtained from the job-specific equations generated by the 23-job and primary (24-job) linkage equations in the holdout and new-job analyses, respectively (i.e., R^2_{cv}). The results appear in Table 8.

In general, the decrease in R^2 associated with using the job-specific linkage equation as compared to the least-squares equation is virtually identical to that expected based on the Browne, Rozeboom, and Lord-Nicholson formulae (i.e., $R^2_{cv} \approx R^2_{adj}$)—the unweighted and weighted (by sample size) average differences ($R^2_{cv} - R^2_{adj}$) being −.007, −.003, .002; and −.014, −.011, and −.008; respectively. In contrast, R^2_{adj} as given by the Wherry formula is typically larger than R^2_{cv} (unweighted and weighted differences of −.019 and −.021, respectively), but this comparison is not particularly appropriate because no population equation exists.

Of the four shrinkage formulae presented in Table 8, the Lord-Nicholson adjustment probably provides the best referent for the holdout analyses (i.e., the 23-job linkage equations), because job-specific linkage equations were generated from a primary linkage equation estimated on a partial sample. The job-specific linkage equations were then applied to a second "sample" (i.e., the holdout job). Thus, the regression parameters for the 23-job equations are not full-sample weights and therefore not the best estimates available. This, in turn, means the parameters for the job-specific linkage equations generated from the 23-job equations are not the best estimates available. Nevertheless, the use of equations containing partial-sample weights rather than full-sample weights suggests that the Lord-Nicholson shrinkage formula provides an appropriate comparison.

For the 7 new jobs that were not part of the original 24-job estimation sample, however, the sample-based linkage equations *were* generated using full-sample weights (i.e., the 24-job primary linkage equation) and used to estimate performance scores for all individuals in a new job (as will be the case upon implementation by manpower planners). Here, one could argue that the Browne formula is the correct referent (i.e., a sample equation based on full-sample weights, applied to a new sample from the population). One might also consider the 24-job equation to be a partial-sample equation, however, given that the data from the new Navy ratings were not incorporated into the sample to yield a 31-job equation. If so, for reasons given above, Lord-Nicholson remains a viable referent.

The conclusion is the same no matter which comparison one chooses: the preponderance of small differences between R^2_{cv} and R^2_{adj} values demonstrates that the linkage methodology provides a means of obtaining predictions of job performance for jobs without criterion data that are nearly as valid (and sometimes more valid) as predictions obtained when (1) criterion data are available for the job, (2) a job-specific least-squares prediction equation is developed, and (3) the equation is applied in subsequent samples.

Comparison of Validity Coefficients to the Literature

Another means of assessing the predictive power of the job-specific linkage equations is to compare their validity coefficients with those reported in the literature for similar predictor/criterion combinations. McCloy (1990) demonstrated that the determinants of relevant variance in performance criteria differ across criterion measurement methods (i.e., written job knowledge tests, hands-on performance tests, and personnel file data and ratings of typical performance), leading to different correlations between a predictor or predictor battery and criteria assessing the same content but measured with different methods. Hence, the most relevant comparisons for the R^2 values given in Table 8 are validity studies involving cognitive ability as a predictor and hands-on measures as performance criteria.

Unfortunately, relatively few studies employ hands-on performance tests as criteria. The vast majority of validity research has used supervisory ratings or measures of training success (e.g., written tests or course grades) as criteria. The preference for these measures is probably due to the ease and lower cost of constructing them, relative to hands-on tests. Nevertheless, there are a few studies that may serve as a standard of comparison.

In a meta-analysis of all criterion-related validity studies published in the *Journal of Applied Psychology* and *Personnel Psychology* from 1964 to 1982, Schmitt et al. (1984) reported the mean correlation between various predictors and hands-on job performance measures to be $r = .40$, based on 24 correlations. They also provided mean validities for specific types of predictors when predicting performance on hands-on tests. General mental ability measures yielded a mean validity $r = .43$ (based on three correlations). Note that meta-analysis corrects the distribution of validity coefficients for range restriction and criterion unreliability.

Hunter (1984, 1985, 1986) reported the correlation between measures of general cognitive ability and hands-on job performance measures to be $r = .75$ in civilian studies and $r = .53$ in the military. These correlations were adjusted for range restriction. A study of military job performance by Vineberg and Joyner (1982) reported an average validity of various predictors for task performance of $r = .31$, based on 18 correlations. In a later study, Maier and Hiatt (1984) reported validities of the ASVAB when predicting hands-on performance tests to range from $r = .56$ to $.59$. Finally, Scribner et al. (1986) obtained a multiple correlation of $r = .45$ when predicting range performance for tankers in the U.S. Army using general cognitive ability (AFQT), experience, and demographic variables.

The R^2 values for the job-specific least-squares and linkage equations given in Table 8 have not been corrected for range restriction or criterion unreliability. For the job-specific least-squares regression equations, values of the multiple correlation range from $r = .26$ (Army MOS 13B) to $r = .71$ (Marine Corps MOS 6113) with unweighted and weighted (by sample size) means of $r = .43$ and $r = .40$, respectively. For the job-specific linkage equations, values range from $r = .18$ (Air Force specialty 272X0) to $r = .68$ (Marine Corps MOS 6112) with unweighted and weighted means of $r = .38$ and $r = .36$, respectively. Clearly, the predictive validity of the job-specific linkage equations lies well within the range of validities that have appeared in the literature.

Summary

Taken together, the results from the cross-validity analyses suggest that the linkage methodology has yielded a performance equation that provides predictions for out-of-sample jobs that are not much below the best one could expect. Predictions are generally better for high-density jobs than for low-density jobs. Nevertheless, the cross-validity analyses have strongly suggested that there is relatively little loss in predictive accuracy when predictions are made for jobs devoid of criterion information. Tempering this finding, however, is the finding that the absolute level of prediction typically ranges from about $R^2 = .10$ to $R^2 = .20$, even when using optimal (i.e., job-specific OLS) prediction equations. Clearly, there remains room for improvement in the prediction of hands-on performance. Nevertheless, from a slightly different perspective, the utility of prediction of job performance for out-of-sample jobs is increased by R^2 percent over what it would be without the primary linkage equation. These results are positive and supportive of the multilevel regression approach to predicting performance for jobs without criterion data.

Discussion

One characteristic shared by validity generalization, synthetic validation, and multilevel regression is that they act as "data multipliers"—they take the results of a set of data and expand their application to other settings when the collection of complete data is too expensive or impossible. Validity generalization does not yield information that is directly applicable to the development of prediction equations for jobs without criteria. Rather, the results suggest (1) whether measures of a particular construct would be valid across situations and (2) whether there is reliable situational variance in the correlations.

Synthetic validity does provide information directly applicable to the task of performance predictions without performance criteria. In fact, as mentioned earlier, no performance criteria of any kind are required. Judgments and good job analytic data alone are sufficient for the production of prediction equations. This would appear to be highly advantageous to small organizations that might otherwise be unable to afford a large-scale performance measurement/validation effort. Furthermore, the largest synthetic validity study ever undertaken, the Army's SYNVAL project, demonstrated these equations to be nearly as predictive as optimal least-squares equations that had been adjusted for shrinkage.

Although not developed for this purpose, multilevel regression analysis has been shown to provide a means of generating equations that occasionally exceed appropriately adjusted validity values from least-squares equations. The results compare favorably with the results from the SYNVAL project, although, unlike the SYNVAL data, the data supplied to the multilevel regression analyses had not been corrected for range restriction. It is possible that the results could be more positive if more appropriate job analytic information were used. Recall that the job characteristic data used in the Linkage project were originally collected on civilian jobs and transferred to the most similar military occupations. A job analysis instrument specifically applied to military jobs might result in better M_j variables and therefore better estimates of the job-specific regression parameters. The Navy has finished a job clustering project that used a job analysis questionnaire developed for Navy jobs—the Job Activities Inventory (JAI; Reynolds et al., 1992) that could easily be modified for application to all military jobs.

One potential drawback of applying multilevel regression techniques is that a number of jobs must have criterion data for estimating the primary linkage equation. The 24 jobs used in the Linkage project did supply enough stability to obtain statistically reliable results based on across-job variation, but including more jobs in the estimation sample would certainly have resulted in better estimates. Increasing the estimation sample should not be unreasonably difficult for larger organizations with some form of performance assessment program in place. For one, the performance criterion does not need to be a hands-on performance test. Written tests of job knowledge or supervisory ratings could serve as criteria just as easily. Performance prediction equations could be developed for new jobs or jobs not having the performance criterion in question.

For example, assessment center research might be helped by this method of estimating predicted performance scores. Sending promising young managers to assessment centers is very costly. A primary equation could be developed based on the individuals who were sent to the assessment centers. Estimated assessment center scores could then be obtained from job-specific regression equations developed from the primary equation. There are a couple of potential drawbacks to this application, including the ability to differentiate between various managerial positions and the effects of range restriction.

The application of multilevel regression techniques also might provide benefits to organizations that are members of larger consortia. The organizational consortium could pool its resources and develop a primary performance prediction equation on a subset of jobs having criterion data across organizations within the consortium. Job-specific equations could then be developed for the remaining jobs.

The research from the Synthetic Validation and Linkage projects has advanced our knowledge of the degree to which performance equations may be created for jobs without criteria. The methodology provided by multilevel regression analysis closely resembles synthetic validation strategies. Both rely heavily on sound job analytic data. After SYNVAL, Mossholder and Arvey's (1984) observation that little work had been done in the area of synthetic validity is no longer true. Further, the Linkage project has demonstrated another successful procedure for generating performance prediction equations that operates without judgments about the validity of individual attributes for various job components. Both procedures should be examined closely in future research because they have the potential for turning an initial investment into substantial cost savings—they make a few data go a long, long way.

ACKNOWLEDGMENTS

The author wishes to thank Larry Hedges and Bengt Muthén for their invaluable help and patience in communicating the details of multilevel regression models and their application, the Committee on Military Enlistment Standards for their challenging comments and creative ideas, Linkage project director Dickie Harris for his support and good humor throughout this research, and the reviewers of the manuscript for their careful reading of a previous version of this chapter. Any errors that remain are the responsibility of the author.

REFERENCES

Browne, M.W.
  1975   Predictive validity of a linear regression equation. *British Journal of Mathematical and Statistical Psychology* 28:79-87.
Campbell, J.P., ed.
  1986   *Improving the Selection, Classification, and Utilization of Army Enlisted Personnel: Annual Report, 1986 Fiscal Year* (Report 813101). Alexandria, Va.: U.S. Army Research Institute.
Campbell, J.P.
  1990   Modeling the performance prediction problem in industrial and organizational psychology. Pp. 687-732 in M.D. Dunnette and L.J. Hough, eds., *Handbook of Industrial and Organizational Psychology*, 2nd ed., Vol. 1. Palo Alto, Calif.: Consulting Psychologists Press.
Campbell, J.P., McCloy, R.A., Oppler, S.H., and Sager, C.E.
  1992   A theory of performance. Pp. 35-70 in N. Schmitt and W.C. Borman, eds., *Personnel Selection in Organizations*. San Francisco, Calif.: Jossey-Bass.
Campbell, J.P., and Zook, L.M., eds.
  1992   *Building and Retaining the Career Force: New Procedures for Accessing and Assigning Army Enlisted Personnel* (ARI Research Note). Alexandria, Va.: U.S. Army Research Institute.
Cattin, P.
  1980   Estimation of the predictive power of a regression model. *Journal of Applied Psychology* 65:407-414.
Crafts, J.L., Szenas, P.L., Chia, W.J., and Pulakos, E.D.
  1988   *A Review of Models and Procedures for Synthetic Validation for Entry-Level Army Jobs* (ARI Research Note 88-107). Alexandria, Va.: U.S. Army Research Institute.
Darlington, R.B.
  1968   Multiple regression in psychological research and practice. *Psychological Bulletin* 69:161-182.

Green, W.H.
1990   *Econometric Methods.* New York: McMillan.
Harris, D.A., McCloy, R.A., Dempsey, J.R., Roth, C., Sackett, P.R., Hedges, L.V., Smith, D.A., and Hogan, P.F.
1991   *Determining the Relationship Between Recruit Characteristics and Job Performance: A Methodology and a Model* (FR-PRD-90-17). Alexandria, Va.: Human Resources Research Organization.
Hedges, L.V.
1988   The meta-analysis of test validity studies: Some new approaches. Pp. 191-212 in H. Wainer and H.I. Braun, eds., *Test Validity.* Hillsdale, N.J.: Erlbaum.
Hollenbeck, J.P., and Whitemer, E.M.
1988   Criterion-related validation for small sample contexts: An integrated approach to synthetic validity. *Journal of Applied Psychology* 73:536-544.
Hunter, J.E.
1984   *The Prediction of Job Performance in the Civilian Sector Using the ASVAB.* Rockville, Md.: Research Applications.
1985   *Differential Validity Across Jobs in the Military.* Rockville, Md.: Research Applications.
1986   Cognitive ability, cognitive aptitudes, job knowledge, and job performance. *Journal of Vocational Behavior* 29:340-362.
Hunter, J.E., and Hunter, R.F.
1984   Validity and utility of alternative predictors of job performance. *Psychological Bulletin* 98(1):72-98.
Knapp, D.J., and Campbell, J.P.
1993   *Building a Joint-Service Classification Research Road Map: Criterion-Related Issues* (FR-PRD-93-11). Alexandria, Va.: Human Resources Research Organization.
Lawshe, C.H.
1952   Employee selection. *Personnel Psychology* 5:31-34.
Laurence, J.H., and Ramsberger, P.F.
1991   *Low Aptitude Men in the Military: Who Profits, Who Pays?* New York: Praeger.
Longford, N.T.
1988   *VARCL Software for Variance Component Analysis of Data with Hierarchically Nested Random Effects (Maximum Likelihood).* Princeton, N.J.: Educational Testing Service.
Lord, F.M.
1950   Efficiency of prediction when a regression equation from one sample is used in a new sample. *Research Bulletin* (50-40), Princeton, N.J.: Educational Testing Service.
Maier, M.H., and Hiatt, C.M.
1984   *An Evaluation of Using Job Performance Tests to Validate ASVAB Qualification Standards* (CNR 89). Alexandria, Va.: Center for Naval Analyses.
McCloy, R.A.
1990   A New Model of Job Performance: An Integration of Measurement, Prediction, and Theory. Unpublished doctoral dissertation, University of Minnesota.
McCloy, R.A., Harris, D.A., Barnes, J.D., Hogan, P.F., Smith, D.A., Clifton, D., and Sola, M.
1992   *Accession Quality, Job Performance, and Cost: A Cost-Performance Tradeoff Model* (FR-PRD-92-11). Alexandria, Va.: Human Resources Research Organization
McCormick, E.J., Jeanneret, P.R., and Mecham, R.C.
1972   A study of job characteristics and job dimensions based on the Position Analysis Questionnaire (PAQ). *Journal of Applied Psychology* 56:347-367.

Mossholder, K.W., and Arvey, R.D.
1984   Synthetic validity: A conceptual and comparative review. *Journal of Applied Psychology* 69:322-333.

Nicholson, G.E.
1960   Prediction in future samples. Pp. 424-427 in I. Olkin et al., eds., *Contribution to Probability and Statistics*. Stanford, Calif.: Stanford University Press.

Primoff, E.S.
1955   *Test Selection by Job Analysis: The J-Coefficient, What It Is, How It Works* (Test Technical Series, No. 20). Washington, D.C.: U.S. Civil Service Commission.

Reynolds, D.H.
1992   Developing prediction procedures and evaluating prediction accuracy without empirical data. In J.P. Campbell, ed., *Building a Joint-Service Research Road Map: Methodological Issues in Selection and Classification* (Draft Interim Report). Alexandria, Va.: Human Resources Research Organization.

Reynolds, D.H., Barnes, J.D., Harris, D.A., and Harris, J.H.
1992   *Analysis and Clustering of Entry-Level Navy Ratings* (FR-PRD-92-20). Alexandria, Va.: Human Resources Research Organization.

Rozeboom, W.W.
1978   The estimation of cross-validated multiple correlation: A clarification. *Psychological Bulletin* 85:1348-1351.

Sackett, P.R., Schmidt, N., Tenopyr, M.L., Kehoe, J., and Zedeck, S.
1985   Commentary on "Forty questions about validity generalization and meta-analysis." *Personnel Psychology* 38:697-798.

Schmidt, F.L., and Hunter, J.E.
1977   Development of a general solution to the problem of validity generalization. *Journal of Applied Psychology* 62:529-540.

Schmidt, F.L., Hunter, J.E., and Pearlman, K.
1981   Task differences as moderators of aptitude test validity in selection: A red herring. *Journal of Applied Psychology* 66:166-185.

Schmidt, F.L., Hunter, J.E., Pearlman, K., and Hirsh, H.R.
1985   Forty questions about validity generalization and meta-analysis. *Personnel Psychology* 38:697-798.

Schmidt, F.L., Hunter, J.E., Pearlman, K., and Shane, G.S.
1979   Further tests of the Schmidt-Hunter Bayesian validity generalization procedure. *Personnel Psychology* 32:257-281.

Schmitt, N., Gooding, R.Z., Noe, R.D., and Kirsch, M.
1984   Meta-analysis of validity studies published between 1964 and 1982 and the investigation of study characteristics. *Personnel Psychology* 37:407-422.

Scribner, B.L., Smith, D.A., Baldwin, R.H., and Phillips, R.L.
1986   Are smart tankers better? AFQT and military productivity. *Armed Forces and Society* 12(2):193-206.

Steadman, E.
1981   *Relationship of Enlistment Standards to Job Performance*. Paper presented at the 1st Annual Conference on Personnel and Training Factors in Systems Effectiveness, San Diego, California.

U.S. Department of Defense
1991   *Joint-Service Efforts to Link Military Enlistment Standards to Job Performance*. Report to the House Committee on Appropriations. Washington, D.C.: Office of the Assistant Secretary of Defense (Force Management and Personnel).

U.S. Department of Labor
1977   *Dictionary of Occupational Titles*. Fourth Edition. Washington, D.C.: U.S. Department of Labor.

Vineberg, R., and Joyner, J.N.
1982   *Prediction of Job Performance: Review of Military Studies.* Alexandria, Va.: Human Resources Research Organization.

Waters, B.K., Barnes, J.D., Foley, P., Steinhaus, S.D., and Brown, D.C.
1988   *Estimating the Reading Skills of Military Applicants: Development of an ASVAB to RGL Conversion Table* (FR-PRD-88-22). Alexandria, Va.: Human Resources Research Organization.

Waters, B.K., Laurence, J.H., and Camara, W.J.
1987   Personnel Enlistment and Classification Procedures in the U.S. Military. Paper prepared for the Committee on the Performance of Military Personnel. Washington, D.C.: National Academy Press.

Wherry, R.J.
1931   A new formula for predicting the shrinkage of the coefficient of multiple correlation. *Annals of Mathematical Statistics* 2:446-457.

Wigdor, A.K., and Green, B.F., Jr., eds.
1991   *Performance Assessment for the Workplace, Volume I.* Committee on the Performance of Military Personnel, Commission on Behavioral and Social Sciences and Education, National Research Council. Washington, D.C.: National Academy Press.

Wing, H., Peterson, N.G., and Hoffman, R.G.
1985   Expert judgments of predictor-criterion validity relationships. Pp. 219-270 in Eaton, N.K., Goer, M.H., Harris, J.H., and Zook, L.M., eds., *Improving the Selection, Classification, and Utilization of Army Enlisted Personnel: Annual Report, 1984 Fiscal Year* (Report 660). Alexandria, Va.: U. S. Army Research Institute.

Wise, L.L., Campbell, J.P., and Arabian, J.M.
1988   The Army synthetic validation project. Pp. 76-85 in B.F. Green, Jr., H. Wing, and A.K. Wigdor, eds., *Linking Military Enlistment Standards to Job Performance: Report of a Workshop.* Committee on the Performance of Military Personnel. Washington, D.C.: National Academy Press.

Wise, L.L., Peterson, N.G., Hoffman, R.G., Campbell, J.P., and Arabian, J.M.
1991   *Army Synthetic Validity Project Report of Phase III Results, Volume I* (Report 922). Alexandria, Va.: U.S. Army Research Institute.

Wright, G.J.
1984   Crosscoding Military and Civilian Occupational Classification Systems. Presented at the 26th Annual Conference of the Military Testing Association, Munich, Federal Republic of Germany.

Part III:
The Cost/Performance
Trade-off Model

As stated in the overview of this report, the JPM Project and the related cost/performance trade-off model grew out of the misnorming of the ASVAB, the congressional interest in the relationship between job performance and recruit quality, and the lack of a clear, coherent framework for establishing recruit quality goals. Throughout the 1980s, the continuing question was: How much quality is enough? With the current level of high quality, the downsizing of the force, and the supply of high-quality recruits exceeding demand, the question now becomes: What level of quality is most cost-effective? The cost/performance trade-off model has been developed as a tool to aid analysts and policy decision makers in answering questions about recruit quality needs and in justifying the costs associated with selected quality mixes. Here we provide a context for the personnel planning process.

Personnel planners in the Services attempt to staff the force structure—divisions, air wings and battle groups, and the supporting infrastructure—with the numbers and types of people necessary to maintain desired readiness levels. As part of this process, the Services must annually recruit numbers of new entrants to the enlisted force to replace those who leave or retire and to reflect planned growth or shrinkage in the overall size of the force. In today's All-Volunteer Force, the military services annually recruit about 200,000 young men and women to become soldiers, sailors, airmen, and marines. These young people receive basic military training and specialized skill training in more than 900 different military occupations. For the military services and the taxpayer, this represents the beginning stages of a significant investment in recruiting resources, training resources, and compensation.

Not all applicants are equally capable of completing basic training or the skill training necessary in the technically demanding jobs constituting an increasing proportion of the modern armed forces. The success achieved by the Services in recruiting the right kinds of young men and women will determine not only how large this investment will be, but also, in part, how effectively the armed forces will meet the challenge of defending the nation's interests. For these reasons, the Services have an incentive to be selective in recruiting only those who are likely to succeed.

For individuals in the youth population, the opportunity to serve one's country is an important part of citizenship. It should not be limited or denied without compelling reasons. Moreover, for many of the nation's youth, service in the armed forces offers the chance to obtain valuable experience and training that could be a major determinant of the potential recruit's future economic well-being. Policies that determine how selective the Services will be in choosing applicants, therefore, are important, and entail a complex array of trade-offs. More selective recruiting policies tend to reduce the effective size of the youth labor market from which the Services may recruit, raising recruiting costs. Presumably, the investment in higher recruiting costs resulting from a more selective policy yields a return in the form of a more capable recruit—one who is more likely to complete training successfully and to perform well on the job. But, because these more selective policies result not only in higher initial recruiting costs, but also deny service opportunities to some willing applicants, the case for more selective recruiting policy should be well grounded in logic. Moreover, the empirical links between higher selectivity, performance, and cost should be well-established.

The Services set their recruit selection policies in terms of applicants' scores on the Armed Forces Qualification Test (AFQT) and an applicant's level of education. The AFQT consists of 4 of the 10 subtests of the Armed Services Vocational Aptitude Battery (ASVAB), the enlistment screening and classification test administered to all applicants. The primary educational criterion is whether the applicant possesses a high school diploma. Those high school graduates scoring in the top half of the distribution on the AFQT (Categories I-IIIA) are considered high-quality applicants.

Recruiting goals are set in terms of the proportion of high-quality recruits desired. A more selective recruiting policy typically means attempting to recruit a larger portion of high-quality applicants. The applicant's score on the AFQT, as well as education status, determines whether the applicant is qualified for the Service. In addition, the Services use various combinations of the ASVAB subtests to form composite scores that are relevant to particular jobs or occupational categories. The applicant's score on the relevant composite must exceed the minimum score established by the Service for that occupation in order to be accepted for training in that particular occupation. Hence, the composite scores determine which jobs or occupational groups the applicant is qualified for. In practice, an individual's AFQT score is highly correlated with most other composite scores, so that the higher-quality applicant, as measured by AFQT, also meets the minimum qualifications for more jobs.

A large number of studies have demonstrated that high-quality recruits—those with above average scores on the AFQT and a high school diploma—perform better in the military, whether performance is measured by training outcomes, job performance tests, speed of promotion, or first-term attrition. But high-quality individuals also cost more to recruit in the volunteer force environment, in which the military must compete with other employers for the services of talented individuals. Therefore, the Services attempt to set recruit quality goals that balance the higher performance and lower attrition costs of high-quality individuals with their increased recruiting costs.

Figure 1 depicts the trade-off between costs and performance that the Department of Defense faces in setting recruit quality goals. As the level of recruit quality increases, both expected military performance and recruiting costs rise. Higher recruiting costs are offset by decreases in attrition-related costs. This figure also illustrates the two empirical linkages that must be established to provide the quantitative measures of the trade-off that are necessary for policy decisions. First, we must know how different levels of recruit quality affect military performance—the top half of the figure. And second, we must understand how changes in average recruit quality affect the components of personnel costs shown in the bottom half.

Understanding these linkages is important for two reasons. First, there should be a solid rationale, grounded in performance and cost differences, for choosing among applicants for military service. To deny a citizen the opportunity to serve his or her country is a serious matter that must be justified with compelling reasons. Second, Congress, as the agent of the taxpayer, has insisted that DoD be able to justify, in terms of increased military performance, the costs of a higher-quality enlisted force. Understanding these linkages is necessary to achieve the maximum return from a declining defense personnel budget.

The first paper in this section, prepared by D. Alton Smith and Paul F. Hogan, provides an overview of the cost/performance trade-off model. The second paper, by Paul F. Hogan and Dickie A. Harris, presents a discussion of the policy and management applications of the model.

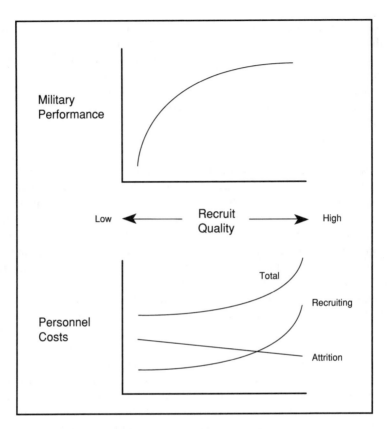

FIGURE 1 Accession quality cost/performance trade-off.

The Accession Quality Cost/Performance Trade-off Model

D. Alton Smith and Paul F. Hogan

INTRODUCTION

This paper describes a model that defines the quantitative linkages between recruit quality, military performance, and personnel costs and uses that information to determine the cost/performance trade-off options available to the Department of Defense (DoD). The model was jointly developed by the Systems Research and Applications (SRA) Corporation and the Human Resources Research Organization (HumRRO) for the Office of the Assistant Secretary of Defense (Personnel and Readiness). It builds directly on the results from the Joint-Service Job Performance Measurement/Enlistment Standards (JPM) Project, which developed and implemented hands-on job performance tests for enlisted personnel.

The Accession Quality Cost/Performance Trade-off Model has many potential applications in accession planning. In particular, it can assist policy makers in:

• Building an accession program, which is the target number of accessions by quality category and occupation group and the associated recruiting resource requirements.

• Evaluating the performance and cost implications of changes in the accession program caused by variation in manpower requirements, recruiting market conditions, or budgets for Service recruiting efforts.

• Efficiently setting job classification standards, the cutoff scores constructed from the Armed Forces Vocational Aptitude Battery (ASVAB) used to screen entrants into particular occupations.

• Assessing the savings potential associated with new processes and tools for selecting and classifying military recruits.

Other papers in this volume describe these applications in more detail and provide examples of the results generated by the model.

This paper is organized as follows. The first section describes the components of the model and identifies some of the strengths and weaknesses of the analytical approach. This is necessarily a summary. More details on the structure of the model and the underlying research can be found in McCloy et al. (1992).

To facilitate its use in policy analysis, we have implemented the analytical model for all four Services in a microcomputer software program, which is described in Human Resources Research Organization et al. (1993). The second section uses the model in two different types of validation tests: comparing model results against actual accession cohorts and varying individual elements in the scenario defining an optimization run. The final section provides a summary of the paper.

DESCRIPTION OF THE MODEL

The structure of the model is shown in Figure 1. It selects, for a set of occupation groups, the number of accessions by recruit quality category that minimizes the sum of recruiting, training, and compensation costs while meeting first-term performance and personnel strength goals. Because the quality-performance and quality-cost relationships vary significantly across the Services, a separate version of the model has been developed for each Service.[1]

It is important to emphasize that the model does not choose the *best* overall recruit quality level, only the cost-minimizing level for a given amount of performance. Selecting the best overall level requires information on the dollar value of performance so that performance gains can be weighed directly against increased costs. In a market setting, in which profit-maximizing firms compete for the services of employees, one can argue that the compensation paid to workers represents the value of their performance. It is much more difficult to make the same claim for public-

---

[1] We are not the first to develop an accession cost/performance trade-off model. The essential structure for such a model is described by Steadman (1981). Cost/performance trade-off models for a small number of Army occupations were implemented and tested by Armor et al. (1982) and Fernandez and Garfinkle (1985). Accession quality models are also currently under development by several of the Services.

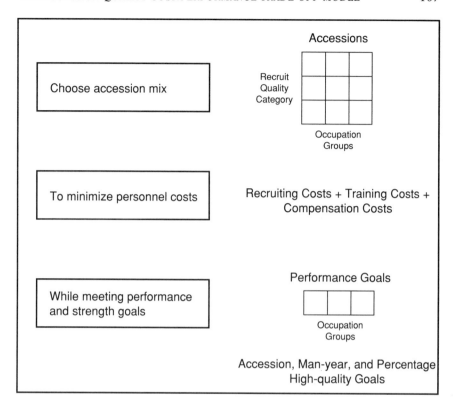

FIGURE 1   Structure of the model.

sector organizations that do not compete. Without using what must be arbitrary performance valuations, the model can be used to define the efficient cost and performance combinations, an improvement in the information currently available to policy makers.[2]

In mathematical terms, the model is a constrained minimization problem with the three elements: accessions by recruit quality category and occupation, which are the variables for which we solve; performance and personnel strength goals, which define the constraints on the problem; and personnel costs, which comprise the objective function. We describe the model in terms of these three elements. An explicit mathematical statement of the model is found in the technical appendix at the end of this paper.

---

[2] For an example of a recruit quality model that employs a performance valuation function, see Nord and Kearl (1990).

Variables: Accessions by Recruit Category and Occupation Group

The model solves for 360 variables for each Service—accessions in 10 recruit quality categories for each of 36 occupation groups.

Recruit Categories

Although there are many ways to categorize recruit quality, our choice should be driven by the policy applications of the model. Therefore, we use the Armed Forces Qualification Test (AFQT) score and high school graduation status, the two characteristics used to establish recruiting goals and measure recruiting performance for DoD. In particular, 10 recruit quality categories are defined by the interaction of 5 standard AFQT score groups (called Category I, II, IIIA, IIIB, and IV) and 2 high school graduation groups (those with high school diplomas and those without).[3]

Additional categories are often used in managing the recruiting process. For example, all Services track enlistments separately by gender because some occupations, by regulation or statute, cannot be filled by women. Although the model would be more useful to accession planners if its recruit categories were also defined by gender, there is little research measuring the linkage between female accession quality and recruiting costs. Without information on both the quality-performance and quality-cost relationships, we cannot add more recruit categories to the model.

Occupation Groups

Both the quality-performance and quality-cost linkages vary with military occupation, which means that the minimum-cost level of recruit quality will also vary by occupation. As an example, consider two occupations with different training costs but the same overall performance goal. The least-cost solution for staffing these occupations will have fewer recruits, of a higher average quality, in the occupation with high training costs. Because of their lower turnover, fewer high-quality recruits have to be trained to generate the same amount of performance, conserving training resources. Given these differences, it is important to understand how high-quality recruits should be allocated across occupations.

Ideally, the aggregate quality requirement for a Service would be determined as the sum of the levels of quality needed in each enlisted occupation

---

[3] Those high school graduates without regular diplomas, such as individuals with General Educational Development (GED) certification, are grouped with the nongraduates, who have similar performance and attrition characteristics.

in that Service. This would yield not only aggregate recruit quality goals but also targets for the process of classifying recruits into occupations. We use grouped, rather than individual, occupations in the model for two practical reasons. First, as described by McCloy in this volume, the performance data used in our model cannot support distinct estimates of the relationship between performance and entry characteristics for every enlisted occupation in DoD. And second, increasing the number of occupations also increases the number of variables in the model. With up to 350 occupations in each Service, finding a solution to the constrained optimization problem becomes time-consuming enough to detract from the model's usefulness as a policy analysis tool.

There are 36 occupation groups in the model, which are defined hierarchically. At the top, occupations are divided into nine groups based on one-digit DoD occupation codes.[4] These codes provide a common classification of occupations across Services, facilitating the definition of performance and strength goals. Each of these nine groups is further divided into four subgroups on the basis of training costs and an index of occupation characteristics.[5] These subgroups were chosen to increase the within-group homogeneity of the occupation groups on variables that are central to the cost-minimization problem, such as training costs. The more homogeneous the groups, the less error is introduced into the recruit quality solution by the aggregation of individual occupations.

Constraints: Performance Goals by Occupation

This is one of the more complex parts of the model. We look first at the definition of performance used in the model, then describe how performance is measured by occupation group and recruit quality category, and finally discuss the setting of performance goals.

Definition of Performance

In the model, performance for an occupation group is the sum of expected hands-on performance test scores over the first term of service for

---

[4] The one-digit DoD occupation codes included in the model are: infantry, gun crews, and seamanship specialists; electronic equipment repairers; communications and intelligence specialists; health care specialists; other technical and allied specialists; functional support and administration specialists; electrical/mechanical equipment repairers; craftsmen; and service/supply handlers.

[5] Specifically, a performance index based only on occupation characteristics is constructed from the performance equations described below. This index indicates how a given individual's performance would vary if assigned to different occupations.

recruits assigned to that group. Specifically, P_i, the performance value for occupation group i, is defined as

$$P_i = \sum_j A_{ij} \left(\sum^t s_{ij}^t \hat{P}_{ij}^t \right)$$ (1)

where

- A_{ij} is the number of accessions from recruit category j into occupation group i. The size of an accession cohort for a Service is equal to the sum of all A's, which, in fiscal 1990, ranged from 90,000 for the Army to 34,000 for the Marine Corps.

- s_{ij}^t is the survival rate to year of service (YOS) t for a category j recruit in occupation group i. Survival rates are the proportion of accessions still in the military at time t; they range from 0 to 1 and decrease with time in service. The model uses a four-year initial term of service to calculate performance values.

- $\hat{P}_{ij}^t$ is the predicted hands-on performance test score at YOS t for a recruit from category j entering occupation group i. The test scores are the percentage of tested task steps completed correctly and can be roughly interpreted as the percentage of the job done correctly. Therefore, the scores can potentially range from 0 to 100. We discuss the approach used to predict these scores below.

The performance value for an occupation group is the result of two calculations. The term in parentheses is the survival-weighted sum of predicted performance test scores over the first term for individuals from a particular recruit category enlisting in that occupation group. As a simple example, assume that AFQT Category I high school graduates enlisting in the Communications and Intelligence group typically stay for four years, making the survival rates all 1, and score 75 on the performance test each year. The expected first-term performance value for these individuals would equal 300.[6] In the second part, occupation group performance is calculated as the weighted sum across recruit categories of these expected scores, with weights equal to the number of accessions from each category.

Occupation group performance is a function of the number and quality distribution of recruits into that group, with quality affecting performance both through the hands-on performance test score and the probability of

---

[6]Specifically, the performance value equals the test score of 75 for first year times the survival rate for that year of 1, plus a score of 75 for second year times the survival rate of 1, etc.

attrition. Although the use of hands-on tests represents an improvement over performance metrics based on training outcomes or job knowledge tests, this measure of occupation performance has four limitations:

• First, it reflects only the ability to perform selected tasks for each job at the entry level. It ignores other dimensions of military job performance, such as leadership potential.

• Second, the same amount of performance for an occupation can be obtained with different combinations of the size and average quality of an accession cohort. As a practical matter, staffing requirements restrict the flexibility to trade off between the number of recruits and their average quality. For example, each tank requires a crew of four to operate efficiently, regardless of the ability of individual crew members.

• Third, equation (1) assumes that the first and last high-quality individual added to an occupation yield the same gain in performance. In fact, the benefit to group performance of that first high-quality recruit is probably greater than the marginal benefit of those that follow.[7]

• Fourth, military performance is, in most cases, a function of both personnel and equipment performance. When force structure (i.e., the number and type of military units) can be varied, the appropriate level of recruit quality will likely be different than that estimated for a fixed force structure.[8]

Fortunately, we can work around some of these limitations by using strength constraints in the model, as described below.

Measuring Performance

Constructing performance values in the model would be straightforward if observations on hands-on performance test scores were available for all occupations, by recruit category and time in service. However, the cost of developing and administering hands-on performance tests limited the data available from the JPM Project. Counting all four Services, we could use results from just 24 occupations to estimate the relationship between recruit

---

[7]For a conceptual discussion of military performance measurement, see Black (1988).

[8] For example, see Daula and Smith (1992). This study estimates the minimum-cost level of recruit quality needed to meet a given level of tank force performance, for which performance is measured by scores on a firing range. Recognizing the role of equipment leads to a higher recruit quality requirement because increasing the quality of tank gunners and commanders allows the same performance requirement to be met with fewer tanks, at a considerable cost savings.

quality and job performance for the model.[9] To generalize from this sample to a set of performance scores for the occupation groups in equation (1), we estimated a multilevel model linking performance scores with characteristics of the test taker and characteristics of the job, as described in this volume by McCloy.[10] For the purposes of the cost/performance trade-off model, the essential advantage of this specification is that the hands-on performance test scores required for equation (1) can be predicted for *any* occupation (or group of occupations) if we know the relevant characteristics of the occupation.

The parameters of the linkage equation were estimated using a sample of approximately 8,400 test scores obtained for the 24 JPM occupations. While there are many ways to evaluate the estimation results, we will focus on two general findings about the AFQT-performance relationship, which is central to the functioning of the model. First, on average, there is a positive, statistically significant correlation between AFQT and job performance. More important, the size of the correlation is roughly consistent with that estimated in other studies using different performance metrics.[11]

Second, our ability to measure the differences in the AFQT-performance relationship across occupations is probably restricted by the small sample of occupations. Although occupations were selected to span the major types of occupations, they turned out not to span the total range of variation in factor scores. As a result, the coefficients that capture the differences in the AFQT-performance relationship across occupations are often not statistically significant and vary with small changes in the functional form of the statistical model. In a later section, we discuss the implications of this finding for the results of the model.

In contrast to the hands-on performance test scores, obtaining survival rates for the performance equation is relatively simple. Longitudinal personnel records, sorted by occupation group and recruit category, can be used to estimate the probability of survival to various time in service points.

---

[9]The results from tests administered in seven additional occupations have been analyzed but not yet incorporated into the model. A preliminary evaluation suggests that adding these jobs will not significantly change the results produced by the model.

[10]Multilevel models are called random coefficients models in the econometrics literature.

[11]This statement is based on comparisons with selected studies familiar to us. For example, the correlation between AFQT and hands-on performance is similar to the correlation between AFQT and tank firing scores in Daula and Smith (1992) and the correlation between AFQT and Army promotion times in Daula et al. (1990). A formal review of the results in the literature should be conducted to validate this point.

Setting Performance Goals

To complete the performance constraints, the model requires the user to specify performance goals for each of the occupation groups. The process of setting these goals raises three issues.

First, in contrast to determining the *number* of enlisted personnel required in each occupation, there are no existing procedures within DoD for estimating personnel performance requirements, using this or any other performance metric. As a result, performance goals for the model have to be set by some sort of expert judgment. In the guidebook developed to assist users of the model, we recommend starting with the calculated performance of a cohort that is generally viewed as having achieved satisfactory performance levels and then making adjustments based on anticipated changes in force structure and performance requirements by occupation group.

Second, as currently structured, the model requires goals for the performance of individual accession cohorts over the first term of their military service. For policy-making purposes, it would be preferable to set a series of *annual* performance goals, which include the contribution of all Service members in the occupation, and have the model derive the series of cohort performance values required to meet the annual goals.

Two problems make this model more difficult to implement than the one outlined here. The job performance test data available from the JPM Project include only tests of first-term personnel within a limited range of years of service. These data are not sufficient to develop the quality-performance relationships that would be required for members of the career force.[12] In addition, we have shown elsewhere that the accession programs generated by this type of model can depend arbitrarily on the number of fiscal years modeled, which is obviously not a desirable characteristic. This problem occurs because, as fiscal years are added to the model, the career lengths used to evaluate accession cohort performance and costs change, leading in most cases to different solutions.[13]

Third, the implementation of the performance constraints does not currently recognize that, because it is generated by a statistical model, the performance for an occupation is really the expected value for an underlying distribution of possible performance values. Strictly speaking, this means that the model chooses the accession mix that meets or exceeds the performance goal 50 percent of the time, which may not be sufficient. It also

---

[12]The lack of career force performance measures does not just prevent us from building an annual model. Even our cohort model might produce different results if performance and costs were measured for the full career, rather than just the first term.

[13]For details see Hogan et al. (1990).

means that the value of improved entry tests, which would reduce the unexplained variance in performance outcomes, cannot be measured in the model. Using performance constraints that recognize the variation in the predicted performance test scores would address both of these issues.[14]

Constraints: Strength Goals

As noted above, we need strength constraints in the model primarily to adjust for the limitations of the performance metric. Three types of strength constraints are included.

First, a total accession constraint ensures that the number of recruits selected by the model fills the number of enlisted vacancies anticipated by a Service. Thus, the model can be prevented from generating solutions that are inconsistent with manning requirements as determined by strength planning models.

Second, a goal can be set for the number of first-term man-years generated by a cohort over the first term of military service. This constraint forces the model to choose an accession cohort that will, in the long run, not alter the desired mix between first-term and career enlisted personnel. Thus, it helps deal with the suboptimization problem that could arise because of the first-term focus of the model.

Third, minimum high-quality percentage constraints (with high quality specifically defined as AFQT I to IIIA, high school diploma graduates) can be placed on the solution by occupation. One adds these constraints to force a solution that recognizes the need for some high-quality recruits in every occupation, based either on their potential as future managers in the occupation or their role in increasing group performance.

Objective Function:
Recruiting, Training, and Compensation Costs

The model minimizes the sum of costs required to recruit, train, and pay an accession cohort over the first term of service. Although the smallest of the three cost categories, recruiting costs have the biggest impact on the structure of the model. In particular, because of the nature of recruiting costs, one must jointly solve for the cost-minimizing level of recruit quality in all occupation groups; a group-by-group approach will not work. The requirement to find a simultaneous solution to recruit quality goals increases

---

[14]Hogan et al. (1990) describes how to implement stochastic performance constraints in the model.

the number of variables in the cost-minimization problem and makes it necessary to estimate the quality-performance linkage across all occupations, a problem already discussed.

Recruiting Costs

Recruiting is characterized by *average* costs that increase with the total number of high-quality individuals recruited. To attract more high-quality individuals into military service, either the monetary incentives for enlisting or the "sales" effort must increase, causing total costs to rise faster than the number of recruits. With increasing average costs, the cost/performance trade-off in occupation A depends not only on the number of high-quality individuals recruited for that occupation, but also on the number recruited for the other occupations. For example, an increase in the number of high-quality individuals recruited for occupation B increases the cost of quality to occupation A, reducing the attractiveness of high-quality individuals in that occupation relative to other quality categories. The increase in quality for occupation B, therefore, leads to a reduction in the cost-effective quality level for occupation A.[15]

In the model, the costs to recruit a particular mix of individuals by quality level are predicted by a cost function. In economics, a cost function describes the minimum costs of producing a particular quantity of output. It is derived from a production function, which shows the relationship between the level of inputs and the output produced, and the prices of those inputs. For recruiting, our production function is based on estimates in the literature of the effect of recruiting resources, such as advertising, and the effect of recruiting market characteristics, such as unemployment, on the number of high-quality enlistments. The depth of the literature supporting the specification of these production functions varies by Service, with the greatest number of studies available for the Army and the least for the Air Force. In the model, we start with what we believe are the best parameter estimates but allow the analyst to modify the parameters of the production function in the software model. The prices of recruiting resources are

---

[15]To understand this point, consider the analytical process of finding the minimum cost level of recruit quality in a single occupation with just two recruit quality categories—high and low. Among the mixes of high- and low-quality recruits that satisfy the performance goal, the model will select that mix for which the cost per unit of performance is the same for high- and low-quality recruits. If these costs are not equal, it is possible to substitute from the high to low cost-per-unit category, providing the same level of occupation performance at a lower cost. Thus, anything that disturbs the balance in per unit costs across recruit quality categories affects the cost-minimizing solution.

calculated from recruiting budgets in a base year. The technical appendix describes the details of the recruiting cost function.

The advantages of determining recruiting costs from the underlying production function are threefold. First, the cost function estimates minimum recruiting costs for a particular mix of enlistments costs, a necessary input in determining the minimum personnel costs for a given set of performance goals.[16] Second, the cost function correctly captures the increasing average cost nature of recruiting costs.[17] This is particularly important given today's recruiting policy issues. Assuming constant average costs would overstate the costs of recruit quality for the smaller enlistment cohorts anticipated in the near future, resulting in recruit quality goals that are too low. Third, cost function estimates of recruiting costs vary with changes in the recruiting market, such as the level of unemployment. This provides the analytical link needed to estimate recruit quality goals for different recruiting environments.

Although the parameters of the underlying production function are selected judgmentally from existing research, the recruiting cost function has been applied to all four Services with reasonable results. For example, the cost function predicts levels of recruiting resource usage and total costs that are within 5 percent of the actual resource usage and costs for a base year. Also, in response to price changes, the cost function adjusts the resource mix as expected to achieve the minimum recruiting costs. For example, when the price of a recruiter increases by 10 percent, total recruiting costs increase by less than 10 percent of the costs associated with recruiters, as more of other recruiting resources are substituted for the relatively more expensive recruiters.

The recruiting cost function in the model is limited, however, by the underlying enlistment supply research in two ways. First, enlistment supply research for most Services has focused primarily on the high-quality group. As a result, it is not possible to implement a recruiting cost function with the 10 quality categories defined above. Instead, we base our cost function on three groups of recruit quality categories, using additional constraints to ensure that the proportion of AFQT I graduates in the high-quality group, for instance, is consistent with historical accession patterns.

Second, our recruiting cost function assumes that the marginal cost of accessing an individual from one of the quality groups does not vary across occupations. This is clearly wrong, because enlistment incentives are usu-

---

[16]A useful by-product of this optimization within the larger optimization problem is the mix of recruiting resources that produces the minimum cost solution.

[17]This makes the objective function a nonlinear function of accessions by recruit category, requiring the use of quadratic programming methods to find the minimum cost solution.

ally targeted on hard-to-fill occupations. Unfortunately, we don't have estimates of the enlistment supply parameters required to efficiently set enlistment bonuses by occupation, as well as relative to other recruiting resources.

Training and Compensation Costs

The derivation of training and compensation costs in the model is much simpler. The average cost of training an individual from a particular recruit category in a given occupation group is determined from two factors. First, we calculate the average cost *per graduate* of basic and initial skill training in the occupation group using Service-supplied estimates of course costs. The initial skill training cost for an occupation group is an accession-weighted average of the initial skill training costs for all occupations in the group. Then, these costs are adjusted by training survival rates for the recruit category to obtain the *per accession* cost of training. Thus, the average training costs within an occupation group are lower for high-quality individuals, who have greater survival rates.

Expected compensation costs over the first term of service are calculated from survival rates and average compensation (basic pay, allowances, and retirement accrual as reported in DoD compensation tables) by year of service.

VALIDATION OF THE MODEL

To facilitate its use in policy evaluation, the model described in the previous section has been implemented in a microcomputer program.[18] Figure 2 summarizes the elements that define a scenario and the results generated by each optimization run.

In this section, we will use the model in two types of validation tests. Setting performance goals equal to the performance expected from the cohorts entering each Service in fiscal 1990, we compare the accession cohorts selected by the model to meet these goals at minimum cost with the actual cohorts accessed. In the second set of tests, we will vary individual elements of the run scenarios and compare the effects of those changes on personnel strength and costs with what would be predicted from theory.

---

[18]David Clifton and Michael Sola of SRA designed and programmed the software. The model is written in FORTRAN and uses routines from the Numerical Algorithm Group (NAG) for the optimization process.

Scenario Elements

- Performance goals, by one-digit occupation group
- Total accessions or man-year constraint
- Minimum high-quality percentages, by one-digit occupation group
- Parameters of the recruit production function
- Inflation factors for recruiting resources, training costs, and military compensation

Optimization Results

- Number of accessions, by quality category and one-digit occupation group
- Number of enlistment contracts, by quality category
- Total performance and average performance per man-year, by one-digit occupation group
- Man-years, by occupation and fiscal year
- Recruiting, training, and compensation costs
- Minimum-cost mix of recruiting resources
- Assumptions and paramenters used in the scenario

FIGURE 2   Model scenario and results.

Model Selections Versus Actual Accession Cohorts

Table 1 compares personnel strengths, percent high-quality accessions, and personnel costs for the actual fiscal 1990 accession cohorts and those selected by the model. To provide a common framework for comparison, "actual" man-years and personnel costs are calculated by applying the survival rates and cost functions in the model to the actual distribution of accessions by quality category and occupation group in fiscal 1990. To generate the model's selected cohort for each Service, we set performance goals for each occupation group equal to the performance expected from the actual fiscal 1990 accession cohort, constrained total man-years to equal the man-years generated by the fiscal 1990 cohort, and established minimums of 40 percent high quality for each occupation group. Thus, the model cohorts are chosen to provide the same performance and man-years as the actual fiscal 1990 cohorts, and we ensure a minimum level of high-quality individuals in each occupation group. Although the overall level of quality selected by the model is always less than the average quality of the actual fiscal 1990 accession cohorts, the differences are not large. Reasonable

TABLE 1 Fiscal 1990 Actuals Versus Model Results

Results	Army		Navy		Marine Corps		Air Force	
	Actual	Model	Actual	Model	Actual	Model	Actual	Model
Personnel strength (Thousands)								
Accessions	84.4	84.0	71.2	70.4	32.1	32.4	34.6	34.4
First-term man-years	243.2	243.2	218.2	218.2	96.0	96.0	110.6	110.6
Percentage high-quality accessions								
Infantry, gun crews, seamanship	57	66	45	40	50	55	82	99
Electronic equipment repair	82	100	78	63	91	45	97	100
Communications, intelligence	73	79	64	55	63	58	93	94
Health care	79	40	59	91	NA	NA	91	99
Other technical specialists	61	77	77	40	87	40	89	97
Support and administration	64	40	60	57	77	43	87	73
Mechanical equipment repair	57	40	54	57	69	63	79	92
Craftsmen	53	40	40	56	69	43	75	91
Service and supply handlers	53	52	23	44	50	70	79	58
All occupations	62	59	54	52	61	55	85	80
Personnel costs (millions)								
Recruiting	632	549	340	313	125	121	167	162
Training	1,539	1,527	1,292	1,255	490	482	376	374
Compensation	4,525	4,530	4,172	4,180	1,837	1,837	2,126	2,127
Total	6,696	6,606	5,803	5,748	2,452	2,439	2,669	2,663

variation in the parameters of the model, such as the enlistment production function, will generate quality levels that exceed the actual levels for each Service. In most models, predicting results close to actual results is an unambiguous validation test. In this model, however, these results are comforting only to the extent that one believes that the current DoD process for setting quality goals, like the model, attempts to find the cost-effective solution.

The table also shows that the models were less successful in replicating the actual quality levels by occupation group. Only 13 out of 35 occupation groups had model results within 10 percentage points of the actual values.[19] There are two potential explanations for the mismatches. First, inaccuracies in measuring either the quality-performance relationship or recruiting costs at the occupation group level, both potential problems in the model, would cause the model to select the wrong quality content. Second, the Services' processes for setting quality goals by occupation, which are not driven solely by performance and cost considerations, may not be producing the cost-effective solution either.

The personnel costs associated with the four cohorts selected by the model are $161 million less than the costs estimated for the actual cohorts. Total recruiting costs are lower because the overall quality of accessions is lower than in the actual cohorts, but training costs also fall because the model allocates high-quality individuals across the occupation groups differently. These savings, while large in absolute value, represent less than 1 percent of the estimated costs of recruiting, training, and paying the actual fiscal 1990 cohorts.

Variation in Model Scenarios

A less ambiguous test of the model is whether the quality levels and associated personnel costs respond as expected to changes in individual elements of the run scenario, such as performance goals. Table 2 shows the results of four variations in the definition of the run scenario for the Navy.[20] In all cases, we will be comparing results with the base case, which shows the Navy cohort selected to minimize personnel costs while meeting fiscal 1990 performance levels.[21]

---

[19]The differences would be more striking if we had not imposed a minimum percentage of high-quality accessions by occupation group.

[20]Qualitatively similar results are obtained for the other Services.

[21]The results here differ from Table 1 because we have not imposed the man-year and minimum high-quality constraints on the solution. It is easier to interpret the results of changing scenario elements when there is only one set of constraints applied.

TABLE 2 Results of Varying Elements of the Model Scenario

Results	Base Case	Performance Reduced 10% (1)	Unemployment Increased (2)	Training Costs Increased 25% (3)	Performance Equals Retention (4)
Personnel strength (Thousands)					
Accessions	70.7	63.4	70.3	70.5	71.0
First-term man-years	218.7	196.1	218.0	218.3	217.8
Percentage high-quality accessions					
All occupations	46.4	50.9	50.8	48.0	24.1
Personnel costs (millions)					
Recruiting	292.2	266.6	291.3	297.6	242.8
Training	1,252.9	1,123.8	1,248.6	1,563.7	1,252.3
Compensation	4,188.4	3,758.6	4,177.7	4,185.0	4,173.6
Total	5,733.5	5,149.0	5,717.6	6,046.4	5,668.7

Scenario 1: Lower Performance Goals

A timely question for recruiting policy is how quality goals should change as the size of military forces decline. For scenario 1, we decrease the performance goals for all occupations by 10 percent from fiscal 1990 levels. As would be expected, fewer accessions are required to meet the new performance goals. Note, however, that the selected cohort is *more* than 10 percent smaller, while the average quality and therefore the average performance per accession has increased. With smaller accession cohorts, the marginal cost of recruiting high-quality individuals decreases. On a performance-to-cost basis, high-quality individuals are now more attractive, and the model selects a higher average quality level for the accession cohort. Mirroring this result, relatively more resources are allocated to recruiting with the lower performance goals, as recruiting budgets decrease by less than the reduction in performance goals.

Scenario 2: A Higher Unemployment Rate

An increase in unemployment reduces the costs of recruiting high-quality individuals. If performance goals remain unchanged, the optimal percentage of high-quality enlistments should increase because these individuals are again more attractive on a performance-to-cost basis. As expected, the cost-minimizing percentage of high-quality enlistments for the Navy increases from 46.4 percent to 50.8 percent when the unemployment rate is increased from the fiscal 1990 level of 5.3 percent to 7.0 percent.[22] To recruit the additional quality, recruiting budgets should remain about the same, even though recruiting is easier with higher unemployment. In other words, the model suggests that the Services should take advantage of the recruiting cost savings available in a slack labor market by recruiting more high-quality individuals.

Scenario 3: Increase in Training Costs

As the costs of training courses increase, it makes sense to select a higher-quality accession cohort because the lower attrition of that cohort conserves training costs. In scenario 3, we increase all training course costs by 25 percent; the remainder of the scenario elements are the same as in the base case. As expected, average quality increases from the base case. Total training costs increase by *less* than 25 percent because of the increased quality of the selected accession cohort.

---

[22]In this scenario, we return to the fiscal 1990 performance goals used in the base case.

Scenario 4: No Effect of Quality on Performance Scores

Suppose that performance scores were not available, so that performance could be measured only by the expected man-years contributed by individuals in different recruit quality categories. Under this definition, the additional performance obtained by recruiting a high-quality individual is less than that when performance includes variation in test scores. As a result, the cost-effective level of quality should fall. To test the contribution of the performance test scores to the determination of recruit quality goals, we ran the model with all performance test scores, the $\hat{P}$'s in equation (1), set equal to 1. For the Navy example, we found that the cost-minimizing level of quality falls by half, to 24.1 percent. Clearly, being able to measure the additional job performance generated by high-quality individuals is important in establishing the correct recruit quality goals.

SUMMARY

Selecting accession quality goals is an important task in defense personnel planning. The chosen level of quality not only affects the average performance that can be expected during the first term of service from cohort members, but it also has a large effect on the future capability of the noncommissioned officer corps, as the military "grows" them from junior enlisted personnel. At the same time, accessing more quality increases the resources that must be devoted to recruiting.

The Accession Quality Cost/Performance Trade-off Model quantifies the linkages between recruit quality, first-term performance, and personnel costs and solves for the level of accession quality by occupation that minimizes the costs of achieving specified levels of performance. Building on the results from previous research into military job performance measurement and enlistment supply, performance equations and recruiting cost functions were estimated and incorporated, along with other information on attrition and training/compensation costs, into a nonlinear optimization model for each of the Services.

Initial tests of the model are promising. It provides quality goals and cost estimates that are generally reasonable, both in terms of historical experience and theoretical expectations. While not without flaws, the model, by quantifying the potential trade-off between performance and cost, should be a useful adjunct to the current processes for determining accession quality goals.

TECHNICAL APPENDIX

This appendix describes the optimization problem underlying the Accession QualityCost/Performance Trade-off Model and outlines the derivation of the recruiting cost function.

The Optimization Problem

Objective Function

We choose the number of accessions, A_{ij}, by occupation group i and recruit category j to minimize the sum of first-term recruiting, training, and compensation costs, given by

$$
R\left(\sum_i \sum_{j \in H} \frac{A_{ij}}{(1-d_j)}, \sum_i \sum_{j \in M} \frac{A_{ij}}{(1-d_j)}, \sum_i \sum_{j \in L} \frac{A_{ij}}{(1-d_j)}, R_p, R_F \right.
$$
$$
+ \left\{ T^B \sum_i \sum_j \left(s_{ij}^B A_{ij} \right) + \sum_i \left[T_i^I \sum_j \left(s_{ij}^I A_{ij} \right) \right] \right\}
$$
$$
\left. + \sum_i \sum_j \left[A_{ij} \sum^t \left(s_{ij}^t C^t \right) \right] \right]
$$

(1)

where

•   The first line shows the recruiting cost function, R. Its arguments include the number of high (H), medium (M), and low (L) quality contracts; the prices of recruiting resources, R_p; and recruiting market factors, R_F. High-quality contracts are the sum of accessions in the AFQT I-IIIA diploma graduate categories, inflated by Delayed Entry Program (DEP) loss rates for each category, d_j. Medium-quality contracts are calculated from accessions in the AFQT I-IIIA nongraduate categories and the corresponding DEP loss rates. Low-quality contracts are computed from accessions and DEP loss rates for the remaining quality categories.

•   The second line shows the training cost calculations. T^B and T^I are the per-graduate variable costs associated with basic and initial skill training (which varies by occupation), respectively. The number of graduates equals the number of accessions times s_{ij}, the survival rate from accession to the completion of basic training (superscript B) and from accession to completion of initial skill training (superscript I).

•   The third line shows compensation costs. Expected first-term compensation for an accession into occupation i from recruit category j is the sum of C^t, the discounted present value of compensation costs in YOS t and the survival rates to each YOS, s_{ij}. Compensation includes basic pay (estimated using average promotion times), allowances, and the retirement accrual charge.

Constraints

These costs are minimized subject to the following set of constraints:

1. Minimum performance values by occupation group, specified as

$$\sum_j \left[A_{ij} \sum^t \left(s_{ij}^t \hat{P}_{ij}^t \right) \right] \geq P_i^* \qquad \text{for all } i \qquad (2)$$

where $\hat{P}_{ij}^t$ is the expected year t performance of a recruit from category j in occupation i and P_i^* is the performance goal for occupation i.

2. Recruit category distribution constraints, which allocate the number of high-, medium-, and low-quality accessions to the 10 underlying recruit categories in proportion to the accession population. These constraints are required because recruiting costs cannot be specified at the detailed recruit category level.

3. Total strength constraints, specified as

$$\text{Accession: } \Sigma_i \Sigma_j A_{ij} \leq \text{ or } \geq A^*$$

$$\text{First-term man-years: } \Sigma_i \Sigma_j A_{ij} \left(\Sigma^t s_{ij}^t \right) \leq \text{ or } \geq MY^* \qquad (3)$$

Accessions and man-years can be constrained to be greater than or less than the goals, A^* and MY^*.

4. High-quality accession constraints, which require a minimum proportion of these accessions in each occupation group.

Only the first two sets of constraints are always active. The strength constraints may be applied at the option of the user.

Optimization Approach

We approximate the recruiting cost function (see the next section) by a quadratic function in high-, medium-, and low-quality accessions. The parameters of the quadratic are estimated by ordinary least-squares regression on a data set generated by the "true" recruiting cost function.

Because recruiting costs are quadratic, we can use quadratic programming techniques to find the cost-minimizing number of accessions by quality categories and occupation group. If, during the optimization process, the trial solution varies significantly from the data used to approximate the recruiting cost function, a new approximation is calculated and the optimization is restarted.

Recruiting Cost Function

The recruiting cost function estimates the minimum costs of recruiting a specified number of individuals in each of the three recruit quality groups defined above. For the recruiting production function, we use results from the enlistment supply literature that describe how recruiting resources, such as recruiters, and market factors, such as unemployment, affect the number of enlistments.

We assume that the number of high-quality contracts signed annually can be described by an enlistment supply function of the form

$$\ln C^H = \ln C_0^H + \alpha_R^H \ln R^H + \alpha_{AD} \ln AD + \alpha_B \ln\left(\frac{B}{v}\right) + \alpha_E \ln\left(\frac{E}{v}\right) + \alpha_F \ln F \quad (4)$$

where C^H is the number of net high-quality contracts signed in a given year; R^H is the number of recruiters "allocated" to the production of high-quality recruits; AD is the amount of advertising; B is the average enlistment bonus paid to high-quality recruits; v is a price index; E is the average cost of education benefits paid to high-quality recruits in recruiting-year dollars; and F represents factors that affect the recruiting market, such as the civilian unemployment rate. The α's are the elasticities of high-quality enlistments with respect to recruiting resources, incentives, and market factors. C_0^H is the constant for the enlistment supply function. (In this equation the superscripts are used as identifiers and not as exponents.)

We assume that the production of medium- and low-cost recruits is limited only by the recruiter effort devoted to testing these individuals and processing them into the military. This implies production functions of the form

$$\ln C^M = \ln C_0^M + \alpha_R^M \ln R^M \quad \text{and} \quad \ln C^L = \ln C_0^L + \alpha_R^L \ln R^L \quad (5)$$

where C^M and C^L are the number of medium-cost and low-quality contracts, respectively. R^M and R^L are the number of recruiters assigned to the medium- and low-cost missions.

The minimum costs associated with recruiting a given number of high-, medium-, and low-quality contracts—C_*^H, C_*^M, and C_*^L—is given by the answer to the constrained minimization problem

$$\text{Min}\left[p^R\left(R^H + R^M + R^L\right) + p^{AD}AD + C_*^H(B+E) + \left(C_*^H + C_*^M + C_*^L\right)T + O\right]$$
$$-\lambda_H\left[C^H - C_*^H\right] - \lambda_M\left[C^M - C_*^M\right] - \lambda_L\left[C^L - C_*^L\right] \tag{6}$$

The first term in brackets represents the recruiting budget; it has five components: the cost of maintaining the production recruiters, the cost of advertising, the expenditures on recruiting incentives, the costs of testing potential recruits, and the fixed costs of recruiting. The fixed costs do not affect the optimal mix of resources but are included in the cost function to provide recognizable budget amounts. The remaining terms in brackets ensure that costs are minimized subject to the constraints of meeting the desired mission.

The first-order conditions for equation (6) describe the solution to the cost minimization problem—the levels of recruiting resources and incentives required to recruit the specified mission at minimum cost. Substituting the first-order conditions into the recruiting budget formula, the first line in equation (6), yields the recruiting cost function

$$\text{Minimum Cost Budget} = \alpha Z\left(C_*^H\right)^{\frac{1+\alpha_B+\alpha_E}{\alpha}} + p^R\left[\left(\frac{C_*^M}{C_0^M}\right)^{\frac{1}{\alpha_R^M}} + \left(\frac{C_*^L}{C_0^L}\right)^{\frac{1}{\alpha_R^L}}\right]$$
$$+\left(C_*^H + C_*^M + C_*^L\right)T + O \tag{7}$$

where

$$Z = \left[\left(C_0^H\right)^{\frac{-1}{\alpha}}(v)^{\frac{\alpha_B+\alpha_E}{\alpha}}\left(\frac{\alpha_R^H}{p^R}\right)^{\frac{-\alpha_R^H}{\alpha}}\left(\frac{\alpha_{AD}}{p^{AD}}\right)^{\frac{-\alpha_{AD}}{\alpha}}(\alpha_B)^{\frac{-\alpha_B}{\alpha}}(\alpha_E)^{\frac{-\alpha_E}{\alpha}}(F)^{\frac{-\alpha_F}{\alpha}}\right] \tag{8}$$

and $\quad \alpha = \alpha_R^H + \alpha_{AD} + \alpha_B + \alpha_E$

Note that the minimum cost budget is a function of four sets of factors: the high-, medium-, and low-contract mission; the parameters of the enlistment supply functions; the prices of recruiting resources; and conditions in the recruiting market.

REFERENCES

Armor, David J., Fernandez, Richard L., Bers, Kathy, and Schwarzbach, Donna
   1982   *Recruit Aptitudes and Army Job Performance.* R-2874-MRAL. Santa Monica, Calif.: The Rand Corporation.
Black, Matthew
   1988   Job performance and military enlistment standards. Pp. 171-198 in Bert F. Green, Jr., Hilda Wing, and Alexandra K. Wigdor, eds., *Linking Military Enlistment Standards to Job Performance.* Committee on the Performance of Military Personnel. Washington, D.C.: National Academy Press.
Daula, T., and Smith, D.A.
   1992   Are high quality personnel cost-effective? The role of equipment costs. *Social Science Quarterly* June:266-275.
Daula, T., Nord, R., and Smith, D.A.
   1990   Inequality in the military: Fact or fiction? *American Sociological Review* October:714-718.
Fernandez, Richard L., and Garfinkle, Jeffrey B.
   1985   *Setting Enlistment Standards and Matching Recruits to Jobs Using Job Performance Criteria.* R-3067-MIL. Santa Monica: The Rand Corporation.
Hogan, Paul F., Harris, Dickie A., Smith, D. Alton, and Clifton, David
   1990   Entry Standards for Military Service: A Cost/Performance Trade-off Model. Working paper presented at the Operations Research Society of America meeting, October 24, 1990.
Human Resources Research Organization, Systems Research and Applications Corporation, and Lewin/VHI, Inc.
   1993   Accession Quality Cost-Performance Tradeoff Model (CPTM) Guidebook. Prepared for the Office of Accession Policy, Assistant Secretary of Defense, Force Management and Personnel.
McCloy, R.A., Harris, D.A., Barnes, J.D., Hogan, P.F., Smith, D.A., Clifton, D., and Sola, M.
   1992   *Accession Quality, Job Performance, and Cost: A Cost/Performance Trade-off Model* (FR-PRD-92-11). Alexandria, Va.: Human Resources Research Organization.
Nord, Roy D., and Kearl, Cyril E.
   1990   *Estimating Cost-Effective Recruiting Missions: A Profit Maximizing Approach.* Alexandria, Va.: U.S. Army Research Institute.
Steadman, Eugene
   1981   Relationship of Enlistment Standards to Job Performance. Office of the Secretary of Defense working paper.

Policy and Management Applications of the Accession Quality Cost/Performance Trade-off Model

Paul F. Hogan and Dickie A. Harris

INTRODUCTION

The ultimate question in setting enlistment standards and recruiting goals, and in programming for recruiting budgets that can achieve those goals, is, "How much quality is enough?" Two other questions, however, must logically precede the answer to this question: "How much does additional recruit quality cost?," and "How much additional performance is generated by higher recruit quality?" The Accession Quality Cost/Performance Trade-off Model attempts to bring together the answers to these two questions, providing policy makers with the information and insights necessary to address the initial question.

In its primary formulation, the model solves for a recruit quality mix that is able to meet desired first-term performance goals, by occupational category, at the lowest cost. It does so by trading the additional recruiting costs for the higher expected performance levels and lower expected attrition costs associated with higher-quality recruits. The model contains three key empirical linkages:

(1) It links recruit quality categories, as defined by AFQT scores, to an empirical measure of actual, "hands-on" performance.

(2) It links not only recruiting costs, but also training costs and compensation costs, to the recruit quality mix chosen.

(3) It links personnel costs, including recruiting, training, and compensation costs, to the costs of generating performance.

Within this framework, a clear, formal definition of an "optimal" recruit quality mix is provided—something that has been absent from much of the debate on enlistment standards and recruit quality. The model solves for a stylized optimal recruit quality mix—one that minimizes the personnel costs of meeting first-term performance goals. From this optimization, recruit quality goals and implied enlistment standards emerge. The recruit quality goals become the best goals in the narrow sense that, within the costs captured in the model, higher- or lower-quality goals would result in greater costs.

In the model, performance is measured as "expected staff-years of performance." This is a combination of expected hands-on performance by occupation and expected retention rate by occupation. A potential recruit's expected performance is defined as the proportion of tasks a first-term enlisted member will have mastered in a given occupational group, as a function of the prospective recruit's characteristics. The relationship between the recruit's characteristics and expected future performance is determined statistically using data from the Joint-Service Job Performance Measurement/Enlistment Standards (JPM) Project (see McCloy, in this volume). Attrition rates, which determine expected staff-years, are also a function of recruit characteristics—most importantly education—and occupation.[1]

Different recruit quality categories, such as high school graduates scoring in Category I on the Armed Forces Qualification Test (AFQT), have different levels of expected performance in the model. Moreover, expected performance can vary across occupational categories. The model chooses recruits from different quality categories to meet occupation-specific performance goals at the lowest possible cost. Hence, there is substitution among recruit quality categories, based on differences between expected performance and cost, and substitution between numbers of recruits and quality of recruits. However, the model assumes that force structure—numbers of divisions, air wings, battle groups, and the equipment, such as tanks, planes and ships—is fixed. There is no substitution, for example, between numbers of tanks and the quality of personnel operating and maintaining tanks. Presumably, such substitution possibilities should be considered when making acquisition decisions. However, once made, the force

---

[1]For a full exposition of the technical details of the model, please refer to Smith and Hogan (in this volume).

structure and its weapons system are fixed, for the purposes of determining recruit quality goals.[2]

The purpose of this paper is to illustrate potential applications of the model in areas of policy, management, and research. By illustrating potential applications, we hope that others in the defense personnel community will be encouraged to use the model and perhaps recommend improvements if it is found useful. An exposition of the model itself can be found in Smith and Hogan (in this volume) and in McCloy et al. (1992).

CONGRESSIONAL DIALOGUE ON
RECRUIT QUALITY STANDARDS

Importance of Linkages

The Services request from Congress and expend recruiting resources to find and enlist "high-quality" recruits who might otherwise not enlist in the Armed Forces. At the same time, the Services choose not to offer enlistment contracts to other "lower-quality" applicants who want to enlist and for which fewer recruiting resources are required. Applicants are screened and divided into the categories based largely on two characteristics—possession of a high school diploma and scores on the Armed Services Vocational Aptitude Battery (ASVAB).

Recruiting goals are set in terms of the proportion of high-quality recruits desired. In general, the higher the proportion of high-quality recruits, the greater the recruiting resources required to achieve the goals, and the larger the numbers of applicants who want to enlist but who are denied. Congress, as an agent of the taxpayers and of the citizens who are applicants for military service, must, in its oversight role, be convinced that enlistment qualification standards and quality goals are fair to the taxpayer and to the applicant. On one hand, to deny a citizen the opportunity to

---

[2]This point has resulted in some confusion. What is fixed and what is variable are important conditions for the analysis of optimal personnel quality. When elements of the force structure, such as tanks, are allowed to vary in the analysis, the effect of personnel quality on the overall effectiveness of the tank should be considered, and substitution between personnel quality and tanks should help to determine the optimal number of tanks, appropriately staffed, to meet given missions at a required level of effectiveness. However, these decisions are not revisited annually. They should be based on expected long-run conditions and should not depend much on factors that affect short-run supply conditions, such as the level of the civilian unemployment rate (and its effect on the relative cost of recruit quality) in a particular year. The types of models used to make the two types of decision—number of tanks in the long run and recruit quality goals this year—are clearly different. Moreover, if the long-run decisions are made correctly, the performance goals for tank crews should be adequately reflected in the model, necessarily a shorter-run model.

serve and defend his or her country because he or she is not "qualified," at the same time allocating more of taxpayers' dollars to achieve the more stringent recruiting goals, is acceptable only if there is a compelling rationale for the quality standards and goals. On the other hand, if quality goals are set too low, the nation and the taxpayer may bear a greater national security risk than they otherwise might, or the taxpayer may be forced to offset the risk through more costly ways of increasing the readiness and capability of the armed forces.

For these reasons, Congress is interested in understanding the method by which qualification standards and recruit quality goals are established. Some assurance is sought that goals and standards strike the right balance among first-term personnel performance, costs, and the interests of the applicant. In particular, Congress has insisted that qualification standards and recruit quality goals be directly linked to the performance or readiness of the first-term force. Over the years, it has been clear that the failure to present to Congress a rigorous method for determining recruit quality goals, based on empirical linkages between enlistment qualification criteria and subsequent job performance, has, during periods of stringent budgets, made recruiting resources and the recruit quality those resources represent a target for reductions. Without a vision of exactly what is sacrificed when recruit quality declines, the case for preserving recruiting budgets is weakened.

The relationship between possession of a high school diploma and the ability of a recruit to persevere and remain in service over the first term is one of the better-established empirical findings in military personnel research. This relationship was established early in the All-Volunteer Force era and was used successfully to screen potential recruits.[3] However, the empirical relationship between ASVAB scores—in particular the subset of the ASVAB constituting the AFQT—and hands-on performance was not well established. Instead, the Services relied largely on the statistical relationship between training success and test scores. Although establishing qualification standards and recruiting goals based on such a relationship is not unreasonable, it is less compelling than a relationship based on actual job performance. In particular, there has been a concern that (a) the relationship between test scores and training success, although important, may arise because the cognitive skills leading to high test scores may be the same cognitive skills that determine success in training, and test scores may be less highly related to actual job performance and (b) spending additional

---

[3]See, for example, Lockman (1978) for one of the earlier studies documenting the relationship between first-term attrition and high school graduation status.

recruiting resources to obtain recruiting goals that are directly linked to first-term performance and readiness is more compelling than goals related to training success.

Role of the Model

The Accession Quality Cost/Performance Trade-off Model can potentially assist in this congressional dialogue regarding quality standards and recruiting goals, how they are formed, the factors affecting them, and the implications of reduction in recruiting budgets in three ways. First, the model provides a logical framework for discussing the determination of recruit quality goals. In the model, quality goals are determined by trade-offs between the greater performance contribution made by higher-quality recruits, the reduced training costs that result, and the higher recruiting costs associated with higher-quality recruits. In general, the model will prescribe higher-quality goals:

- the greater the performance differences between high-quality recruits and other recruits;
- the higher the costs of training; and
- the lower the recruiting costs of high-quality recruits compared with other recruits, at the margin.

The model provides systematic predictions, consistent with this framework, for how recruiting quality goals and recruiting resources should change when total demand for accessions change, the unemployment rate increases, or there is a shift in the proportion of jobs for which higher-quality recruits have a significant performance advantage or for which training costs are high. These consistent relationships provide a logical foundation for articulating to Congress why the Services' recruit quality goals are changing and why additional recruiting resources may be justified.

Second, recruit quality is linked empirically to a measure of hands-on job performance. When reductions in recruiting budgets are proposed, the ensuing discussion of potential effects may proceed beyond the discussion of input quality—AFQT scores and education—to more output-related measures such as a decline in an index of the proportion of job tasks that the lower-quality recruit cohort is likely to master. Moreover, the quality standards and recruit quality goals are potentially linked to a measure closer to predicted job performance. Presumably, this serves to provide a more compelling rationale in terms of personnel readiness, one that is easier for Congress and its constituents to accept.

Third, the formulation links not only recruiting resources, but also training resources and compensation costs, to the determination of recruit quality goals. This linkage should shift congressional interest from a narrow focus on the recruiting budget to the broader implications of recruit quality for training costs and other personnel costs. Saving an additional $5 million in the recruiting budget by lowering recruit quality standards will be more difficult to justify if it results in an additional $7 million in training costs.

We present below several examples of how the model may be useful in discussions with Congress concerning the recruiting budget.

Hypothetical Accession Program

Potentially the most important role that the model can fulfill is that of providing a rational framework for discussing recruit quality and recruiting budget issues with Congress. If used in this role, the model will provide a basis for asserting (a) why a particular recruit quality distribution is best, (b) why the recruiting budget is needed to buy that particular recruit quality mix, and (c) the consequences, for the performance and readiness of the first-term enlisted force, of failing to obtain the recruiting budget necessary to purchase the desired level of recruit quality.

Consider the following hypothetical example of the presentation of an Army recruiting program to the relevant subcommittees in Congress, for fiscal 199X (Table 1).

The hypothetical Army fiscal 199X recruiting program (represented by the Proposed Program row in the table) represents approximately a 20-percent reduction in accessions relative to the fiscal 1990 program. The detailed output from the model (not shown) indicates an exception for the military occupational specialties (MOS) in the electronics equipment repairers occupational category.[4] Accessions for these occupations remain at about the fiscal 1990 level, reflecting a shift toward a somewhat more technologically intensive force. The proposed program calls for an accession plan of 65.6 percent high-quality recruits (AFQT Category I-IIIA high school graduates) and almost 100 percent high school graduates. It is the least costly way to achieve the first-term performance goals for this recruiting cohort and produces an expected average level of performance per staff-year over the first term of service of almost 57 percent.[5]

Alternative cases 1 and 2 show how the proposed program would change if the economic scenario, represented by the unemployment rate, were to

---

[4]These include MOS 26Y, 27E, 35G, 35H, 39D, 39G, 31V, 35E, 35R, 68R, 24C, 24G, 24H, 24J, 24K, 27B, 27F, 27G, 27N, 45G, 93D, and 31E.

[5]The 57 percent is an index value that under one interpretation suggests that the typical recruit will perform about 57 percent of the tasks in his or her occupation over a 4-year term at enlistment.

TABLE 1 Fiscal 199X Hypothetical Army Recruiting Program

	Market Assumption (Unemployment)	Accessions	High-Quality Recruits	High School Graduates	Performance/ Staff-Yr.[a]	Recruiting Budget	Total Cost
Proposed program	6.5%	67,500	65.6%	99.2%	56.9%	$357.2M	$5,258.8M
Alternative case 1	7%	67,500	66.9%	99.2%	57%	$344.5M	$5,240.7M
Alternative case 2	5.3%	67,500	64.9%	99.3%	56.7%	$414.3M	$5,319.5M

[a]Performance per staff-year is total performance (which will equal the performance goals) divided by the total number of staff-years generated by the selected cohort over the first term of service.

vary. The performance goals remain unchanged, but the least costly mix of recruits to achieve those goals will change because the unemployment rate affects the relative cost of recruiting high-quality and lower-quality recruits. Case 1 shows that if the unemployment rate were 7 percent, the best program would include a higher level of high-quality recruits, and total costs would fall. Case 2 indicates that, at a lower unemployment rate, the best recruiting plan includes a slightly lower mix of high-quality recruits and a slightly higher total cost.

Question: Why is the proportion of high-quality recruits in the proposed accession plan higher than in the fiscal 1990 plan?

Answer: This plan allows us to achieve overall performance goals at the lowest cost. There are three reasons why the plan is somewhat more high-quality-intensive than the fiscal 1990 plan. First, with the reduction in overall accession demand from about 84,000 to about 67,500, the relative cost of high-quality recruits has declined. Therefore, it is less costly to achieve our goals by recruiting a slightly greater portion of high-quality recruits. Second, the unemployment rate in fiscal 199X is expected to be about 6.5 percent, compared with a rate of about 5.3 percent in fiscal 1990, a fact that also tends to reduce the cost of higher-quality recruits. Third, although the fiscal 199X plan is approximately a 20-percent reduction from fiscal 1990 accession levels, accessions in the electronics equipment repair category have remained at about the fiscal 1990 level. The training costs in this highly technical area are the highest in the Army, and the differences in expected performance between higher- and lower-scoring recruits in these areas are relatively large. Hence, we find that higher-quality personnel in this category reduce the costs of achieving the overall performance goals. For example, while the average expected performance per staff-year is about 57 percent overall, in the electronic equipment repair category it is 63 percent—the highest among the nine DoD occupational categories. (Note that the average expected performance per staff-year in the fiscal 1990 recruit cohort is about 56 percent). Since the proportion of recruits entering this occupational group increases, it is efficient (i.e., less costly) to meet performance goals by recruiting a slightly higher-quality mix.

Question: What are the consequences of reducing the proposed recruiting budget by $25 million?

Answer: A $25 million reduction in the recruiting budget would result in a failure to meet our performance or readiness goals for this cohort over the first term of service, or would force us to meet those goals in a more costly way. Overall first-term performance would decline by about 0.5 percent, the proportion of high-quality recruits would fall to about 59 percent, and the average expected performance per staff-year would decrease to about 56.3 percent.

The logical structure of the model provides a relatively solid basis for discussing why a particular accession program is, or is not, reasonable and the consequences of changing that program. The recruit quality level prescribed by the model is that which produces the specified level of first-term expected performance, as measured by the statistical relationship between recruit characteristics and hands-on performance, at the lowest cost to the taxpayer. Clearly the empirical measures and parameters in the model can be improved. But, perhaps for the first time, this framework, if generally accepted, will reduce the issues regarding the recruiting budget to two fundamental points: Is additional first-term performance worth its additional costs? Can we improve on the estimates of the empirical relationships underlying the framework?

The logical framework of the current model suggests the following propositions:

• Changes in the recruiting budget result in changes in the quality mix of recruits, which, in turn, affects the expected performance of the first-term enlisted force.

• The optimal quality mix depends, inter alia, on the state of the recruiting market, training costs, the distribution of openings among jobs, and the differences in expected performance among recruit quality categories.

These propositions and others are illustrated by the preceding example, by the examples that follow in this section, and by the section on program development and evaluation. Although estimates of certain empirical relationships can clearly be improved over time, the general framework of this model provides clear and reasonably convincing insights to the question of why a particular recruiting program is, or is not, the preferred program.

Proposed Recruiting Budget Reduction[6]

Consider a hypothetical case in which Congress is contemplating a reduction in the Navy's fiscal 1995 recruiting budget (Table 2). The requested recruiting budget for fiscal 1995 is estimated to result in 68,066 accessions, of which 51 percent are high-quality recruits (AFQT Category I-IIIA high school graduates) and 84.5 percent are high school diploma

---

[6]This example is taken from Human Resources Research Organization et al. (1993).

TABLE 2 Implications of a Hypothetical Reduction in the Navy
Recruiting Budget

	Fiscal 1996 Navy Case	12% Reduction in Recruiting Budget
Recruiting costs	$345.1M	$302.4M
Accessions	68,066	68,066
Staff years	216,813	216,813
Percent high quality	51%	38.9%
Expected performance per staff-year	61%	55.3%

graduates. Approximately 216,813 staff-years of service are expected to be
produced by this accession cohort over the first term of service. The requested recruiting budget, derived from an application of the model, is $345
million for this program. The required budget, which is sensitive to the
competitiveness of the recruiting market, was estimated under the assumption that the civilian unemployment rate would be 6.3 percent and the ratio
of starting military pay to pay in the civilian youth labor market would be
0.89. The average expected performance per staff-year of service in this
program would be 61.0. That is, based on the statistical relationship between recruit quality characteristics and expected performance over the first
term of service, a typical recruit from this cohort will, on average over the
first term of service, have been proficient in about 61.7 percent of the tasks
required by the job, under one interpretation of this performance measure.
In the early part of the term, proficiency or job performance will be lower
than this, and it will be higher than this in the latter part of the term.

Given the intense competition for funding, Congress considers a 12-
percent reduction from an already lean recruiting budget. However, autho-
rized end strength is unchanged, so that approximately 68,066 accessions
and 216,813 staff-years of service will still be required from this cohort
over the first term of service. To obtain the same number of accessions
with a 12-percent reduction in the recruiting budget, the percentage of high-
quality recruits will decline from 51.1 to 38.9 percent in this hypothetical
example. In the traditional analyses, this is as much as can be said. Because the model is directly linked to performance, however, we can say that
the average performance per staff-year declines by about 10 percent, from
61 to about 55.3 percent. Under one interpretation of the performance
metric, this means that the typical recruit from this cohort will be approxi-

mately 10 percent less proficient in the performance of his or her assigned tasks over the first term of service. Since high school graduation status is the main predictor of first-term attrition, the model selects only high school graduates in order to keep the same number of staff-years in both recruiting cases. In the case of reduced recruiting resources, the model selects lower-aptitude applicants from the high school graduate pool.

Cost of Performance Increase

As the force structure declines over the next few years, the readiness level of the remaining force will become increasingly important. Two key questions may become: What is the price of increasing the expected performance of the first-term force? and Should all budgets, including recruiting, decline proportionately? We provide answers to these questions at two levels of accession and performance (Table 3). The first is at the level of accessions that entered the Army in fiscal 1990. In all cases, we assume a civilian unemployment rate of 6.3 percent. The total cost for exercising the model to obtain the minimum cost mix of recruits to achieve the performance level expected for that accession cohort was approximately $6,496.2 million. This includes recruiting, training, and compensation costs of an entry cohort of about 84,900 accessions, providing approximately 245,049 staff-years of service over the first term, with an average expected performance per staff-year of about 56.2 percent. Using the model again to estimate the cost of an accession cohort with a performance level that is 1 percent above this cohort's level, the total cost rises to $6,562.1 million. The additional cost of a 1 percent increase in performance, in this example, is about $65.9 million.

Now, consider the cost of providing a level of first-term performance that is about 20 percent less than that of the fiscal 1990 accession cohort.

TABLE-3   Cost of First-Term Performance

	Fiscal 1990 Performance Level		Fiscal 1990 Level less 20%	
	—	+1%	—	+1%
Total cost	$6,496.2M	$6,562.1M	$5,179.6M	$5,245.4M
Incremental cost of performance		+65.9M		+$65.8M
Percent change in total cost			−21%	
Recruiting cost	$395.6M	$398.4M	$336.4M	$338.4M
Percent change in recruiting cost			−15%	

This cohort will represent about 67,200 accessions and provide 195,000 staff-years of service. The total cost of this reduced accession cohort is estimated to be about $5,179.6 million. This is about 21 percent lower than the fiscal 1990 cohort. If we were to increase the performance level by the same increment (approximately 1 percent) as in the case of the fiscal 1990 cohort, the total cost, including recruiting, training, and compensation costs, would rise to about $5,245.5 million. The incremental cost of additional performance at the lower level of total performance is about the same, at $65.8 million. Hence, the model indicates that, to a first approximation, the incremental price of what can be interpreted as a measure of personnel readiness in the first-term force is not likely to become significantly less costly with downsizing.

Importantly, however, although the model indicates that total costs fall approximately in proportion to the overall change in the performance level of the first-term force in the Army, optimal recruiting budgets fall less than proportionately, according to the model's prescriptions. In moving from the level of performance implied by the fiscal 1990 Army recruiting cohort to a level of performance that is approximately 20 percent less, the recruiting budget declines from $395.6 million to $338.4 million, a decline of only 15 percent. This is because, as we have illustrated elsewhere, the effect of downsizing is to reduce the relative cost of achieving performance goals through high-quality recruits.[7]

THE MODEL AS A PROGRAM DEVELOPMENT AND EVALUATION TOOL

Evaluation of a Recruiting Program

The model may help provide answers to four general questions raised by a Service's accession plan. The first question concerns *feasibility*. Is it likely that the program can be achieved with the resources programmed? Is the budget too low or too high, given the recruiting goals? The second question concerns the program's *efficiency*. Here, there are two aspects of efficiency: (a) Do the recruiting resources programmed represent the most efficient combination of recruiting resources (recruiters, bonuses, advertising, and education incentives) to achieve programmed recruit quality goals? and (b) Is the programmed recruit quality mix the most efficient way to achieve the performance goals for the accession cohort over the first term of service? The third question is *robustness*. What are the consequences of changes in the economic environment or of cuts in recruiting resources? And fourth, the model may provide assistance in answering, perhaps, the

---

[7]See Smith and Hogan, (in this volume), for a more detailed discussion of this point.

TABLE 4   Hypothetical Army Accession Program, Fiscal 199Y

	Accessions	High-Quality Recruits	High School Graduates	Staff-Years	Recruiting Costs	Total Costs
Fiscal 199Y program	67,522	62.2%	95.3%	194,571	$401.2M	$5,234.1M

most difficult question of all: Has the *right level of first-term performance* been programmed? How much more (less) costly are increases (decreases) in performance goals, and should they be changed?

Feasibility

Consider the following hypothetical accession program submitted by the Army for fiscal 199Y (Table 4).

Can this program be executed with the recruiting resources programmed? The economic assumptions for the recruiting market in fiscal 199Y are that the unemployment rate will be about 6.5 percent and the ratio of military starting pay to wages in the civilian youth labor market will be about 0.89. Given these assumptions, the recruiting cost function component of the model indicates that a recruiting budget of about $383.6 million will be necessary if the Army is to achieve a 62.2 percent high-quality mix for 67,522 accessions in the projected fiscal 199Y recruiting market. The recruiting cost function estimates the minimum cost of achieving a given accession plan, assuming an efficient use of recruiting resources. Moreover, it is a rough check on feasibility. That the Army budget request of $401.2 million is within 5 percent of the budget prescribed by the model suggests that there is not an important program execution issue. The model suggests that the Army has programmed about the right level of resources for its plan.

Efficiency

The recruiting resources programmed by the Army for fiscal 199Y are shown in Table 5. The only major difference between the recruiting resource mix programmed by the Army and the mix prescribed by the recruiting cost function is that the Army appears to have programmed too much advertising and too few recruiters. Costs can be reduced by cutting advertising by about $7 million and adding about 200 recruiters. Enlistment

TABLE 5   Mix of Recruiting Resources

	Recruiters	Advertising	Enlistment Bonus (average)	Education Benefits (average)
Army program	3,522	$24.6M	$300	$200
Recruiting cost function	3,722	$15.6M	$443	$190

bonuses could also be raised, but an increase is not likely to be approved by Congress.

From the model, we can calculate the optimal recruit quality mix necessary to achieve the level of performance expected from the accession cohort programmed by the Army. This optimal mix is the one that achieves the same expected performance goal as the Army has programmed, but at the minimum recruiting, training, and compensation cost. In Table 6, the two programs are compared.

According to the model, the overall recruit quality mix that the Army has programmed is about right. However, the model's solution suggests that the same level of expected performance over the first term of service can be obtained from an accession cohort that is approximately 525 accessions less and about $60 million less in recruiting costs. The reason the model is apparently able to do as much with less is that it distributes high-quality recruits among the various occupational categories much differently than does the Army program. However, the model's allocation is based purely on classification efficiency criteria—it does not take into account the preferences of recruits, for example—and considers only measured performance and cost differences in the allocation. Moreover, it reduces the total number of accessions. Without additional constraints, for example, the model does not recognize that a minimal proportion of higher-quality re-

TABLE 6   Comparison of Fiscal 199Y Program to the Model Optimum

	Accessions	High-Quality Recruits	High School Graduates	Staff-Years	Recruiting Costs	Total Costs
Fiscal 199Y program	67,522	62.2%	95.3%	194,571	$401.2M	$5,234.1M
Model optimum	66,987	61.5%	99.1%	194,265	$338.2M	$5,173.2M

cruits might be necessary in some occupations simply to provide leadership, and that individuals are necessary to fill specific positions.

Robustness

How sensitive is the recruiting program to the underlying assumptions? The recruiting cost function portion of the model can be used to analyze the effect of changes in the economic environment on required recruiting budgets. For example, the recruiting budget is based on the assumption that the civilian unemployment rate in fiscal 199Y will be about 6.5 percent and the ratio of military pay to the pay of civilian youth will remain at about 0.89. If, instead, the economy were to grow at a faster rate and the civilian unemployment rate were to decline to 6 percent, approximately $20 million more in recruiting dollars would be required to execute the program. If, in addition, first-term pay were frozen for the next year while the nominal wages of civilian youth were to increase by about 4 percent, the relative military pay ratio would decline to about 0.86. These two factors together would mean that an additional $32 million dollars would be necessary to execute the program, compared with the recruiting market conditions that are assumed in the program. Note, also, that with changes in the recruiting environment, the optimal recruiting program would itself change. This also could be analyzed by the model.

Level of Performance

How much more would an across-the-board 1 percentage point increase in aggregate first-term performance cost, relative to that underlying the program? How much would be saved by a 1 percent reduction? The model was used to estimate these trade-offs, with the *performance level* implied by the Army fiscal 199Y program as a baseline.[8] (Note that in this analysis, the model is run unconstrained so that accessions and staff-years are permitted to vary. If both are held constant when the performance goals vary, the trade-offs become harsher, because all the change in performance must be accomplished through changes in the recruit quality mix.) The cost/performance trade-off around the hypothetical Army fiscal 199Y program is

---

[8]The case used as the program is not the Army's program, but the model's cost-minimizing solution to achieve the same level of performance implied by the program. Since the solutions to the changes in performance around the program level of performance are cost-minimizing optimums, comparing these to the actual program, and not the cost-minimizing alternative, would confuse the incremental cost of performance changes with the savings from the cost-minimizing solution to the fiscal 199Y program.

TABLE 7   Cost-Performance Trade-off Around the Army, Fiscal 199Y Program

Performance Level	Accessions	High-Quality Recruits	Recruiting Cost	Total Cost	Change Relative to Program
Program	67,002	62.6%	$342.1M	$5,178.5M	—
+1%	67,854	62.4%	$347.0M	$5,244.5M	+$66M
−1%	66,105	63.5%	$339.3M	$5,112.5M	−$66M

shown in Table 7. The table illustrates that the incremental cost of a percentage change in performance, around the hypothetical fiscal 199Y Army program, is about $66 million. The change in recruiting costs is relatively small, in this case because the model was permitted to change accessions and staff-years, as well as recruit quality, to meet the changes in the performance goal. The changes in the recruiting budget would have been much larger if accessions and/or staff-years were held constant in the analysis.

Developing the Service's Accession Programs

The model can be a potentially useful tool in developing an accession program, as well as evaluating a program. We illustrate its use as a development tool using a hypothetical Navy accession program.

Setting Recruit Quality Goals and Resources

In building the Navy accession program for fiscal 199Z, the initial guidance is that the overall level of first-term performance should be about that of the actual fiscal 1990 accession cohort, except that it should be about 20 percent smaller. Starting with the expected first-term performance implied by the fiscal 1990 program but reduced by 20 percent, the model is exercised to achieve the following program for fiscal 199Z (Table 8).

The model is run unconstrained (case 1) to produce the recruiting program shown in the table. This unconstrained case, according to the model, is the minimum cost program that produces a level of expected performance over the first term of service that is about 20 percent less, across occupation, than the actual fiscal 1990 program. It assumes that the fiscal 199Z recruiting environment will include an unemployment rate of 6.5 percent and that the ratio of Navy starting pay to civilian youth wages will be 0.89. With 90 percent high school graduates and almost 61 percent high-quality recruits, this appears to be a solid first start.

TABLE 8   Hypothetical Navy Fiscal 199Z Accession Plan

	Accessions	High-Quality Recruits	High School Graduates	Expected Performance/ Staff-Year	Recruiting Cost	Total Cost
Case 1	55,753	60.7%	90%	60.7%	$240.4M	$4,554.0M

TABLE 9   Adding an Accession Constraint

	Accessions	High-Quality Recruits	High School Graduates	Expected Performance/ Staff-Year	Recruiting Cost	Total Cost
Case 1	55,753	60.7%	90%	60.7%	$240.4M	$4,554.0M
Case 2	56,000	59.1%	89%	60.6%	$237.7M	$4,554.3M

The Navy strength planner notes, however, that in her opinion there are too few accessions. At least 56,000 accessions are required to meet strength goals, according to the Navy's Force Analysis Simulation Technique model—the personnel inventory model used by the strength planners. The model is rerun, this time imposing the constraint that accessions equal 56,000. The results are shown in Table 9.

Compared with the unconstrained case, the second case results in slightly lower quality and a slightly higher total cost. However, the differences are small and recruiting costs are actually lower in case 2.[9] As a result, it is decided that the plan shown as case 2 will be brought to the Admiral for approval.

Adjusting for Specific Occupational Groups

Upon briefing the Admiral, it is discovered that a planning error was made. Although most occupations will decline by 20 percent relative to the fiscal 1990 accession level, the need for nuclear power skills will decline by 30 percent, reflecting the retirement of a significant number of Los Angeles-class attack submarines and several ballistic missile submarines.[10] Most

---

[9]Because in the first run the model was permitted to satisfy the performance goal without an explicit constraint on accessions, the total cost must be less than case 2 or the model would be in error.

[10]This is purely hypothetical.

TABLE 10   Fewer Nuclear-Trained Accessions

	Accessions	High-Quality Recruits	High School Graduates	Expected Performance/ Staff-Year	Recruiting Cost	Total Cost
Case 1	55,753	60.7%	90%	60.7%	$240.4M	$4,554.0M
Case 2	56,000	59.1%	89%	60.6%	$237.7M	$4,554.3M
Case 3	55,163	61.3%	89.8%	61.5%	$238.3M	$4,502M

of the nuclear-trained petty officers are in the electronics repair DoD occupational group. The solution, therefore, is to adjust the performance goals in this occupational group downward by 30, instead of 20 percent. The strength planner's accession goals also did not consider this additional reduction. Case 3 is a rerun reflecting this change—it does not include an explicit accession constraint (Table 10).

Case 3 comes out well. Recruit quality, as a percentage, is higher, and expected performance per staff-year increases to 61.5 percent. The recruiting budget goes up, slightly, even though total accessions and required performance go down, because high-quality recruits become somewhat less costly at the slightly lower level of accessions. Total costs of case 3 are less than in those for cases 1 and 2.

Sensitivity to Recruiting Market Conditions

Upon review of the accession plan, analysts from the Chief of Naval Recruiting Command suggest that the recruiting market assumptions may be too optimistic from the viewpoint of recruiting. They note that a pay cap is planned and that the economy's growth rate is picking up. To test the

TABLE 11   Sensitivity to Recruiting Market Assumptions

	Accessions	High-Quality Recruits	High School Graduates	Expected Performance/ Staff-Year	Recruiting Cost	Total Cost
Case 1	55,753	60.7%	90%	60.7%	$240.4M	$4,554.0M
Case 2	56,000	59.1%	89%	60.6%	$237.7M	$4,554.3M
Case 3	55,163	61.3%	89.8%	61.5%	$238.3M	$4,502M
Case 4	55,375	58.2%	89.9%	61.3%	238.8M	$4,510.9M

sensitivity of the plan to market assumptions, the plan is rerun assuming that the unemployment rate falls to 6 percent in fiscal 199Z and that the ratio of Navy starting pay to the youth wage falls to 0.86. The results are shown in the Table 11 as case 4.

According to the model, the optimal adjustment to a tighter recruiting market includes reducing the high-quality goal somewhat, with only a modest increase in recruiting costs. Hence, although not desirable, it appears that a tighter recruiting market, as defined by this scenario, would not have disastrous consequences for the accession plan.

SETTING ENLISTMENT STANDARDS: ANALYSIS OF OCCUPATION-LEVEL ENLISTMENT STANDARDS

Thus far, the applications of the model have focused primarily on higher-level program decisions—the overall recruit quality mix, recruiting resources, the nature of the trade-off between budgets and performance, and the effects of the environment on these quantities. In these areas, the model is potentially helpful in formulating recruit quality goals and the resources necessary to achieve those goals, and in understanding the linkages between recruit quality goals and measures of job performance. Underlying the formulation of overall recruit quality goals, however, are enlistment standards by occupation category or group. In this section, we explore potential uses of the model in helping to understand the implications of occupation-level enlistment standards and changes in those standards and, perhaps, in providing insights concerning setting and evaluating alternative enlistment standards at the occupation level.

Minimum standards for entry into particular jobs or occupational categories are established by the Services on the basis of selected composites or combinations of the 10 subtests of the ASVAB. The subtests selected to form a composite score for a particular job or occupational category concern measurement of those abilities that are considered to be the most relevant for predicting training success and future performance in that occupation. Typically, a composite will consist of a linear combination of 2 to 5 of the 10 ASVAB subtests. Table 12 shows the combinations of subtests used by the Services.

For each military occupation or groups of occupations, the Services set a cutoff score for the relevant composite. To enter that occupation or occupational group, recruits must have achieved at least the minimum score on the relevant composite.[11]

---

[11]In addition, the Services attempt to achieve a degree of classification efficiency. That is, they attempt to channel recruits into those occupations in which the recruit not merely quali-

TABLE 12   Service Aptitude Composites, Fiscal 1990

ASVAB Subtests	Army Composites	Navy Composites	Marine Corps	Air Force
Mechanical Comprehension (MC)		General Technical	General Technical	General (AR+WK+PC)
Arithmetic Reasoning (AR)	Electrical (GS+AR+MK+EI)	Electrical	Electrical	Electrical
Numerical Operations (NO)	Clerical (PC+WK+AR+MK)	Clerical	Clerical	Administrative (NO+CS+WK +PC)
Math Knowledge (MK)	Mech/Maint (NO+AS+MC+EI)	Mechanical (AS+MC++ WK+PC)	Mech/Maint (AR+AS+MC+EI)	Mechanical (GS+2AS+MC)
General Science (GS)	Combat (AR+CS+AS+MC)	Basic Electronics (GS+AR+2MK)		
Paragraph Comprehension (PC)	Field Artillery (AR+CS+MK+MC)	Engine/Boiler/ Machine (AS+MK)		
Auto/Shop (AS)	Operations/Food (NO+AS+MC+ WK+PC)	Mechanical Repair (AR+AS+MC)		
Electronics Information (EI)	Surveillance/ Communications (AR+AS+MC+ PC+WK)	Submarine (AR+MC+PC+WK)		
Word Knowledge (WK)	Skilled Technical (GS+MK+MC+ PC+WK)	Communications Technician (AR+NO+CS+WK+PC)		
Coding Speed (CS)	General Maintenance (GS+MK+EI+AS)	Hospital (GS+MK+PC+WK)		

The model does not include the 10 ASVAB subtests, or the composites built from those subtests, as variables or dimensions in the model. How-

fies, but appears especially well qualified based on his or her composite scores. Actual classification decisions also take into consideration the current staffing of the occupation, the convening dates for training classes, and the recruit's preferences, so that the matches that are made are less than perfect from the more narrow criterion of classification efficiency.

ever, constraints can be set in the model on the minimum proportion of high-quality recruits in each of the occupational groups, but a floor on the minimum AFQT score in the occupational group cannot be directly imposed. Ostensibly, then, it would be difficult to use the model to analyze occupation-specific, minimum composite scores or enlistment standards. However, the model does use the AFQT as the overall measure of aptitude and recruit quality and the AFQT is itself a composite score consisting of the mathematical and verbal subtests: paragraph comprehension (PC), word knowledge (WK), arithmetic reasoning (AR), and mathematics knowledge (MK). Moreover, the individual subtest scores on the ASVAB are positively correlated, some of them highly so. Hence, in general, for a given composite, the higher the cutoff score, the higher the average quality of recruits entering that occupation, as measured by the AFQT.

We are able to analyze the effect of job-specific enlistment standards, defined by minimum cutoff scores on relevant composites, by approximating the effect that the cutoff scores would have on the minimum proportion of high-quality recruits entering that occupational group. We do this by establishing two links: a link between applicant's composite scores and the applicant's AFQT score and a link between the AFQT-equivalent minimum cutoff score of an occupation and the minimum proportion of high-quality recruits in that occupation. The first relationship is based on a strong correlation, in most cases, between an applicant's AFQT score (which itself is the linear combination of four ASVAB subtests) and composite scores. We can predict, with a reasonable degree of accuracy, the AFQT-equivalent of the various composite scores. The second is based on a statistical relationship between the AFQT-equivalent minimum score and the observed proportion of high-quality recruits in that occupation. Hence, we set enlistment standards in terms of a minimum percentage of high-quality recruits that, in turn, is statistically related to the underlying composite scores.

It is the relationship between the composite cutoff score and an estimate of overall high-quality recruits, as measured by the AFQT, that we wish to explore using the model. To do this, we use the overall correlation between AFQT and the various composite scores to construct an AFQT-equivalent of each composite score. We then use the observed relationship between the AFQT-equivalent cutoff score for the occupational category and the proportion of high-quality recruits in the occupation to analyze enlistment standards and the implications of changing those standards. Recognizing throughout that the relationship between underlying occupational enlistment standards, as measured by cutoff scores on the composites, and the proportion of high-quality recruits in the occupational category is less than perfectly precise, the types of analyses that can be conducted using the model include:

- Comparisons of the proportion of high-quality recruits in each occupational category prescribed by the model to the proportion that occurs under current enlistment standards, under otherwise similar conditions;
- The implications for overall recruit quality goals and for the costs of imposing a minimum proportion of high-quality recruits implied by current enlistment standards; and
- The implication for overall recruit quality goals and for the costs of raising (or lowering) enlistment standards in one or more occupational categories.

We describe below how the model can be used to provide insights into the implications of occupation-specific enlistment standards. In particular, we can compare the costs of an unconstrained solution—one with no occupation-specific minimum standards—to a solution in which the equivalent of the current, occupation-specific enlistment standards are imposed, the costs associated with raising (or lowering) one or more cutoff scores, and the likely distribution effects of inefficiently high standards (according to the model) in some occupational groups.

The Empirical Linkages:
Linking Occupational Enlistment Standards to High Quality[12]

Two types of equations will connect occupation-level enlistment standards, as measured by composite cutoff scores, to the proportion of high-quality recruits in an occupational category. These are a set of equations relating AFQT to composite scores and an equation relating the proportion of high-quality recruits (% HQ) in an occupation to the cutoff score:

$$\text{AFQT}_i = a + b\text{Composite}_{ij} + e_{ij} \tag{1}$$

and

$$\%\text{HQ}_j = a' + \text{AFQTCut}_j + e'_j \tag{2}$$

The first equation is estimated using data from individual applicants and is a regression of an individual i's AFQT score on that individual's score on composite j. The second equation, estimated at a more aggregate level, establishes the relationship between the cutoff score in an occupational category as its predicted AFQT-equivalent and the percentage of high-quality recruits (AFQT Category I-IIIA high school graduates) in the occupa-

---

[12]This section provides a brief overview of some of the underlying empirical relationships developed to link the model to enlistment standards.

TABLE 13   Regression of AFQT Scores on Composites with Goodness of Fit As Measured by R^2

Army		Navy		Marine Corps		Air Force	
Composite	R^2	Composite	R^2	Composite	R^2	Composite	R^2
General technical	.94	General technical	.94	General technical	.88	General technical	.97
Electrical	.89	Electrical	.89	Electrical	.89	Electrical	.90
Clerical	.98	Clerical	.65	Clerical	.85	Administrative	.66
Mech/maint	.61	Mechanical	.65	.Mechanical	.65	Mechanical	.49
Combat	.72	Basic electronics	.93				
Field artillery	.86	BT/EN/MM	.72				
Operations/food	.72	Mech rep.	.65				
Surveill/comm	.77	Submarine	.88				
General maint	.74	Comm. tech.	.80				
Skilled tech.	.89	Hospital	.92				

tional category.   The AFQT cutoff score is obtained by using the first equation to translate the composite score minimum to an AFQT score.

In general, the relation between AFQT and composite score is quite strong, as measured by R^2, for most composites, as illustrated in Table 13.

Occupations in the model are aggregated into nine groups corresponding to one-digit DoD occupation codes.   Using the relationship estimated between composite scores and AFQT, we compute the AFQT equivalent of the minimum cutoff score for each DoD occupation code.   This is done by computing the weighted average of the AFQT-equivalent cutoff score for the Service occupations within a DoD occupation category, in which the weights are the proportions of recruits entering that occupation in fiscal 1990.   Table 14 shows the AFQT-equivalent minimum score required for each DoD occupation category using this method.

Finally, occupation-level quality constraints in the model (the equivalent of minimum enlistment standards) are not in terms of AFQT scores, but rather are in the form of the proportion of high-quality recruits—the proportion of AFQT Category I-IIIA high school graduates.   Moreover, one of the important outcomes of enlistment standards by occupational group is the proportion of high-quality recruits that enters that occupational category, partly as a result of the standard.   For this reason, we developed an equation predicting the proportion of high-quality recruits that enter an occupation as a function of the AFQT-equivalent cutoff score.

A regression equation describing the relationship between the weighted-average AFQT minimum cutoff score in an occupation and the proportion of high-quality recruits in that occupation is estimated by using the actual

TABLE 14   AFQT-Equivalent Average Enlistment Standards by DoD
Occupation Code Average Implied AFQT Cutoff Scores

DoD Occupation Code	Army	Navy	Marine Corps	Air Force
(0) Infantry, gun crews	32.13	39.08	29.32	32.13
(1) Electronic equip. rep.	55.29	62.96	67.23	63.3
(2) Comm/intel	43.45	51.97	45.45	49.88
(3) Medical/dental	44.87	44.22	——	45.22
(4) Other technical	45.56	55.51	48.01	46.64
(5) Functional support/admin.	43.18	61.13	54.06	45.78
(6) Elect/mech. repairers	46.48	45.22	45.88	44.65
(7) Craftsmen	38.60	50.14	38.23	37.89
(8) Service/supply handlers	40.71	38.78	31.84	35.36

TABLE 15   Proportion of High-Quality Recruits as a
Function of AFQT-Equivalent Enlistment Standards

Service	Intercept	AFQT Cutoff Score Coefficient	R^2
Army	12.7	1.18	.49
Navy	−19.9	1.51	.57
Marine Corps	19.4	1.11	.78

proportion of high-quality recruits by occupational category in fiscal 1990.
For three of the Services, the proportion of high-quality recruits is regressed
on the estimated AFQT-equivalent cutoff score for fiscal 1990.[13]   The results are shown in Table 15.   Interpreted literally, these equations imply that
a one-unit increase in the average enlistment standard of an occupational
category, as measured by the AFQT equivalent, is associated with a 1.18
percentage point increase in the proportion of high-quality recruits assigned
to that category in the Army, a 1.51 percentage point increase in the Navy,
and a 1.11 percentage point increase in the Marine Corps.   Using these
equations, we can estimate the percentage of high-quality recruits implied
by the AFQT cutoff scores in effect in fiscal 1990 and compare this per-

---

[13]The Air Force is omitted because fiscal 1990 data were not available to us at the time of
this writing.

TABLE 16   Percentage of High Quality by Occupational Category, Fiscal 1990

Occupation Code	Army		Navy		Marine Corps	
	Actual	Predicted	Actual	Predicted	Actual	Predicted
0	56.8	50.6	45.2	39.1	50.0	51.9
1	81.8	77.9	78.4	75.1	90.9	94.0
2	72.7	63.9	63.8	58.5	62.5	69.8
3	78.5	65.6	58.8	46.8		——
4	61.1	66.4	76.5	63.9	87.1	72.6
5	63.5	63.6	59.8	72.4	76.9	79.4
6	56.7	67.5	53.5	48.3	68.9	70.3
7	53.1	58.2	39.7	55.8	68.6	61.8
8	52.6	60.7	22.6	38.6	49.6	54.7

centage to the actual percentage of high-quality recruits in that category. This comparison is shown in Table 16.

Applying the Cost/Performance Trade-off Model to Enlistment Standards

It should be emphasized that the model was not specifically designed for the micro analysis of occupation-specific enlistment standards, but for more programmatic issues involving aggregate recruit quality goals and recruiting resources. Hence, we recommend that the particular applications of the model that follow be considered experimental in nature, with a goal of examining the potential of the model in this direction and, if promising, the modifications that would be necessary to realize this potential.

The performance goals and performance equations in the model do not capture all of the factors that influence readiness and job performance. Enlistment standards, represented by externally imposed, minimum cutoff scores for entry into an occupation, may capture factors affecting performance and readiness that are not included in the model. But there does not literally exist a composite cutoff score such that all those who score below that score will be unsuccessful, while all those above that score will be successful. Instead, the trade-offs are between probabilities of success or higher levels of performance and higher recruiting costs associated with more restrictive entry standards. In principle, enlistment standards by occupation should be determined by weighing the expected costs and the anticipated benefits. The model can highlight some of the cost implications of enlistment standards and changes in those standards and possibly help managers improve the way in which standards are set and revised.

Effect of Current Enlistment Standards
on Recruit Quality and Cost

In this example, we compare an optimal fiscal 1990 Army accession plan when there are no explicit enlistment standards by occupation—the unconstrained case—with the same scenario when actual enlistment standards are approximated by the implied minimum proportion of high-quality recruits shown in the "Predicted" column for the Army in Table 16. These proportions are an estimate of the effect of the fiscal 1990 minimum composite scores, discussed above, translated into minimums in terms of high quality.

First, the model is run without placing any constraints on the proportion of high-quality recruits within the occupational categories. This is equivalent to having no explicit enlistment standard except that implied by the cost-minimizing solution. Next, accessions are constrained to the actual fiscal 1990 level of 84,400 for the Army. Both the performance goals and the accession goals are held constant in the analysis. The average expected performance per staff-year in fiscal 1990 was 56 percent. First-term staff-years, however, are unconstrained. In the unconstrained case, there are no minimum high-quality percentages set by occupational group.

The alternative case imposes the fiscal 1990 enlistment standards on each occupational category. The case is precisely the same as the unconstrained case except that constraints on the minimum proportion of high-quality recruits that enter each of nine occupation groups are set to equal the percentages in the "Predicted" column from Table 16 for the Army. These minimums are the implied representation of minimum composite scores in the model. A comparison of the two cases is shown in Table 17.

The results indicate that the imposition of this approximation to enlistment standards raises the proportion of high-quality recruits accessed by about 10 percentage points, from 50.9 percent in the unconstrained optimum to 60.7 percent in the enlistment standards case. The unconstrained case also suggests that higher quality is warranted in combat arms (infantry, gun crews, and seamanship), electronic equipment repairers, and communication and intelligence, compared with the allocation in the enlistment standards case, with lower portions of quality recruits allocated to the other occupational groups. One might presume that if it were optimal to have a high-quality proportion that exceeds the minimum enlistment standard, the allocation would be made. This is not necessarily the case, as the results indicate. The required quality allocation due to the enlistment standards in some occupations raises the marginal cost of high-quality recruits to all other occupations. Hence, while the unconstrained case suggests that 74 percent high-quality recruits is optimal for combat arms, when minimums are imposed in other occupations, the optimal, given the allocation to other

TABLE 17 Effect of Enlistment Standards on Recruit Quality and Cost: Army Example

Occupations	Unconstrained Case	Fiscal 1990 Enlistment Standards
Percentage High-Quality Recruits		
Infantry, gun crew, seamanship	74.0	50.6
Electronic equipment repair	100	77.9
Communications, intelligence	100	64.0
Health care	0	65.6
Other technical	60.4	66.5
Support and administration	22.6	63.6
Mechanical equipment repair	0	67.5
Craftsmen	0	58.2
Service and supply handlers	20.1	58.2
All	50.9	60.7
Costs		
Recruiting	$ 482.2M	$ 578.3M
Training	1,532.9M	1,534.7M
Total	6,573.0M	6,645.8M

occupations, falls to 50.6 percent, which is just sufficient to satisfy the minimum.

Reducing Enlistment Standards

What are the effects of reducing the AFQT-equivalent, minimum enlistment standard by, say, 10 AFQT percentile points? Using the statistical relationship between AFQT-equivalent enlistment standards and the proportion of high-quality recruits in an occupation, we can solve for the new, lower enlistment standards implied by the reduction. These lower standards are entered into the model to establish a third case. The results of running this case are shown in Table 18, which includes the previous two cases for comparison. With the reduction in enlistment standards there is a concomitant reduction in costs and the proportion of high-quality recruits accessed. Recruiting costs fall by about $72 million, compared with the case in which the original standards are imposed, and the proportion of high-quality recruits falls by about 7 percentage points. Interestingly, the proportion of high-quality recruits in the occupations in which the unconstrained case suggests high quality is most efficient—electronic repair and communications and intelligence—rises. This occurs because the marginal cost of high-quality recruits to those occupations is reduced as high-quality recruits

TABLE 18   Reduction in Enlistment Standards

Occupation	Unconstrained Case	Fiscal 1990 Enlistment Stamdards	10 AFQT-Equivalent Point Reduction in Fiscal 1990 Enlistment Standards
Percentage High-Quality Recruits			
Infantry, gun crew, seamanship	74.0%	50.6%	38.8%
Electronic equipment repair	100%	77.9%	100.0%
Communications, intelligence	100%	64.0%	71.7%
Health care	0%	65.6%	53.8%
Other technical	60.4%	66.5%	58.6%
Support and administration	22.6%	63.6%	51.8%
Mechanical equipment repair	0%	67.5%	55.7%
Craftsmen	0%	58.2%	46.4%
Service and supply handlers	20.1%	58.2%	48.9%
All	50.9%	60.7%	53.1%
Costs			
Recruiting	$482.2M	$578.3M	$501.9M
Training	$1,532.9M	$1,534.7M	$1,533.5M
Total	$6,573.0M	$6,645.8M	$6,591.1M

are released from occupations in which enlistment standards are inefficiently high, according to the model. An implication of these results is that inefficiently high enlistment standards in one or more occupations raises the cost of high-quality recruits to other occupations, potentially reducing the proportion of high-quality recruits in those occupations below what would have been optimal in the unconstrained case.

SUMMARY

We have examined some potential applications of the Accession Quality Cost/Performance Trade-off Model from several different perspectives:

•   As a vehicle for explaining to Congress the rationale for an accession program and its relation to personnel readiness over the first term of service;
•   As a tool for evaluating a program;
•   As a model for developing an accession program; and
•   As an aid in setting enlistment standards.

Its actual contribution to the defense community in any of these areas depends on the willingness of analysts and researchers to work with the model. The purpose of this paper is to suggest, by examples, that the effort may be worth the cost. We believe that the model, as it currently exists, provides an excellent logical framework for analyzing recruiting and first-term performance issues, perhaps providing new insight. Most important, the model may be helpful in articulating the reasons for programmatic choices regarding recruiting goals, resources, and first-term performance.

The model clearly can be improved, particularly by refining some of the underlying empirical relationships. Continued use of the model, however, is a necessary condition for improvement. Only through its use will its strengths and weaknesses be shown and its value be demonstrated.

REFERENCES

Human Resources Research Organization, Systems Research and Applications Corporation, and Lewin/VHI, Inc.
  1993   *Accession Quality Cost-Performance Tradeoff Model (CPTM) Guidebook.* Prepared for the Office of Accession Policy, Assistant Secretary of Defense, Force Management and Personnel.
Lockman, Robert F.
  1978   A Model for Estimating Premature Losses. In R.V.L. Cooper, ed., *Defense Manpower Policy: Presentations from the 1976 Rand Conference on Defense Manpower.* R-2396-ARPA. Santa Monica, Calif.: The Rand Corporation.
McCloy, R.A. Harris, D.A., Barnes, J., Hogan, P.F., Smith, D.A. Clifton, D., and Sola, M.
  1992   Accession Quality, Job Performance, and Cost: A Cost/Performance Trade-off Model" Report No. FR-PRD-92-11. Alexandria, Va.: Human Resources Research Organization.

Workshop Participants
June 9-11, 1993

COMMITTEE ON MILITARY ENLISTMENT STANDARDS

Bert F. Green, Jr. (Chair), Department of Psychology, Johns Hopkins
University
David J. Armor, Institute of Public Policy, George Mason University
Frank P. Brechling, Department of Economics, University of Maryland
Bengt Muthén, Graduate School of Education, University of California,
Los Angeles
Charles R. Roll, Jr., The Rand Corporation, Washington, D.C.
Richard J. Shavelson, Graduate School of Education, University of
California, Santa Barbara

OFFICE OF THE ASSISTANT SECRETARY OF DEFENSE

W.S. Sellman, Director for Accession Policy (Personnel and Readiness)
Jane M. Arabian, Assistant Director for Enlistment Standards (Personnel
and Readiness)
William J. Carr, Manpower Policy Analysis (Personnel and Readiness)

CONTRACTORS' REPRESENTATIVES

Jeffrey Barnes, Human Resources Research Organizaton
Dickie A. Harris, Human Resources Research Organization

Rodney A. McCloy, Human Resources Research Organization
Paul F. Hogan, Lewin/VHI Inc.
D. Alton Smith, Systems Research and Applications Corporation
Mike Sola, Systems Research and Applications Corporation

AIR FORCE REPRESENTATIVES

Capt. Sandra C. Beveridge, HQ Recruiting Services
Lt.Col. William Cummings, HQ USAF
George Germadnik, HQ Recruiting Services
Larry Looper, Armstrong Laboratory
Capt. Gary Macomber, Armstrong Laboratory
Col. William Strickland, Armstrong Laboratory
Maj. Robert Treat, HQ USAF

ARMY REPRESENTATIVES

LTC Thomas Daula, U.S. Military Academy
Peter Greenston, U.S. Army Research Institute for the Behavioral and
    Social Sciences
MAJ John Hershberger, Army Recruiting Command
Peter Legree, U.S. Army Research Institute for the Behavioral and Social
    Sciences
MAJ James Lewis, Research and Plans Division
Abraham Nelson, U.S. Army Research Institute for the Behavioral and
    Social Sciences
LTC Barry Scribner, U.S. Military Academy
LTC David A. Thomas, U.S. Military Academy
MAJ James Thomas, Deputy Chief of Staff, Personnel

MARINE CORPS REPRESENTATIVES

Neil Carey, Center for Naval Analyses
Adebayo Adedeji, Center for Naval Analyses
James Marsh, Headquarters, Marine Corps

NAVY REPRESENTATIVES

Ronald Beardon, Navy Personnel Research and Development Center
Gerald Laabs, Navy Personnel Research and Development Center
John E. Leather, Office of the Chief of Naval Operations
Clessen Martin, Recruiting Plans and Programs
CAPT G.S. McInchok, Recruiting and Retention Programs
Edward J. Schmitz, Research and Studies

OTHER GUESTS

V. Jon Bentz, Sears Roebuck, Retired
Lawrence Hanser, RAND Corporation
Charles R. Hoshaw, Headquarters, USN, Retired
Robert Tinney, Defense Manpower Data Center-East
Thomas Ulrich, Defense Manpower Data Center-East
Harold P. Van Cott, National Research Council, Retired
Lauress L. Wise, Defense Manpower Data Center-West

NATIONAL RESEARCH COUNCIL STAFF

Alexandra K. Wigdor, Division on Education, Labor, and Human
    Performance
Anne S. Mavor, Committee on Military Enlistment Standards
Carolyn J. Sax, Division on Education, Labor, and Human Performance